Revolution and Pedagogy

REVOLUTION AND PEDAGOGY

INTERDISCIPLINARY AND TRANSNATIONAL PERSPECTIVES ON EDUCATIONAL FOUNDATIONS

Edited by

E. Thomas Ewing

REVOLUTION AND PEDAGOGY
© E. Thomas Ewing, 2005.

First published in 2005 by
PALGRAVE MACMILLAN™
175 Fifth Avenue, New York, N.Y. 10010 and
Houndmills, Basingstoke, Hampshire, England RG21 6XS
Companies and representatives throughout the world.

PALGRAVE MACMILLAN is the global academic imprint of the Palgrave Macmillan division of St. Martin's Press, LLC and of Palgrave Macmillan Ltd. Macmillan® is a registered trademark in the United States, United Kingdom and other countries. Palgrave is a registered trademark in the European Union and other countries.

ISBN 1–4039–6920–5

Library of Congress Cataloging-in-Publication Data

Revolution & pedagogy: interdisciplinary and transnational perspectives on educational foundations / edited by E. Thomas Ewing.
    p. cm.
"The conference from which this collection of essays derives, Revolution and Pedagogy: Interdisciplinary Perspectives on Change, was convened at the Ohio State University, Columbus, Ohio, April 18–20, 2002"—Acknowledgements.
    Includes bibliographical references and index.
    ISBN 1–4039–6920–5
    1. Critical pedagogy—Congresses. 2. Education and state—Congresses.
3. Social justice—Congresses. I. Title: Revolution and pedagogy.
II. Ewing, E. Thomas, 1965–

LC196.R47 2005
370.11′5—dc22   2004058730

A catalogue record for this book is available from the British Library.

Design by Newgen Imaging Systems (P) Ltd., Chennai, India.

First edition: May 2005

10 9 8 7 6 5 4 3 2 1

Printed in the United States of America.

CONTENTS

ACKNOWLEDGMENTS

The conference from which this collection of essays derives, *Revolution and Pedagogy: Interdisciplinary Perspectives on Change*, was convened at the Ohio State University, Columbus, Ohio, April 18–20, 2002. It was proposed as a project to the Center for Middle Eastern Studies (Dr. Alam Payind, Director) and received initial logistical and material support from that Center and from the Center for Slavic and East European Studies (Prof. Halina Stephan, Director), as part of the Centers' respective U.S. Federal Title VI-funded programming. Subsequently, a major grant from Ohio State University's Mershon Center (Prof. Ned Lebow, Director) and additional contributions from the Department of Near Eastern Languages and Cultures supported the conference and this publication.

Besides the essays published here, additional presentations at the conference were by Shelley Wong (Ohio State University) on strategies and ideologies of teaching English in the People's Republic of China; Sharofat Mamadambarova (Khorog State University, Tajikistan) and Sunatullo Jonboboev (Tajik Technical University, Dushanbe, Tajikistan) on the Aga Khan Humanities Project and post-Soviet reforms in Central Asian pedagogy; Eva-Marie Stolberg (University of Bonn) on Bolsheviks and experimental pedagogy in the Soviet Union; Brian M. Puaca (University of North Carolina) on postwar German Exchange Scholars in the United States; Murat Ozturk (Yale University) and Ertan Aydin (Cankaya University) on the pedagogy of the People's Houses in revolutionary Turkey; and Hyunjing Shin (University of Hawaii) on critical/radical pedagogy in Confucian environments, specifically in Korea.

The convener of the Ohio State conference wishes to thank the members of the program committee, Shelley Wong, Yucel Demirer, and Saba Boland, for their assistance, the sponsors for their material support, all the presenters and audience at the conference for their vigorous oral commentary on the delivered papers, and especially Tom Ewing for his energetic editorial supervision of this volume for publication. The editor wishes to thank Margaret Mills for organizing the conference and initiating the effort to publish these chapters, Amy Nelson for her advice and encouragement, the Mershon Center at the Ohio State University for its continuing support, and all the contributors for taking time out from their many other commitments to continue working on these chapters.

Margaret A. Mills, Conference Convener
E. Thomas Ewing, Editor

Notes on Contributors

Cati Coe received her Ph.D. from the University of Pennsylvania, is an Assistant Professor of Anthropology at Rutgers University, Camden, and has published *The Dilemmas of Culture in African Schools* (2005) as well as articles in *Journal of American Folklore, Field Methods*, and *Africa Today*.

Roland Sintos Coloma received his Ph.D. from the Ohio State University, is an Assistant Professor in Education at Otterbein College, and has published articles in the *Journal of Curriculum Theorizing* and the *International Journal of Sexuality and Gender Studies*.

Yücel Demirer received his Ph.D. from the Ohio State University, and has published articles on Turkish culture and politics.

E. Thomas Ewing received his Ph.D. from the University of Michigan, is an Associate Professor in the Department of History at Virginia Tech, and has published *The Teachers of Stalinism* (2002) as well as articles in *Gender & History, Russian Review*, and *The Journal of Women's History*.

Margaret A. Mills received her Ph.D. from Harvard University, is a Professor in the Department of Near Eastern Languages and Cultures at the Ohio State University, and has published *Rhetorics and Politics in Afghan Traditional Storytelling* (1993) as well as numerous books and articles.

Esmail Nashif received his Ph.D. from the University of Texas, is an Assistant Professor in the Department of Sociology and Anthropology at Bir Zeit University, and has published *Hadith. Short Stories* (1998) as well as articles in Palestinian journals and the edited collection *Palestinians since Oslo*.

Barak A. Salmoni received his Ph.D. from Harvard University, is an Assistant Professor in National Security Affairs at the U.S. Naval Postgraduate School, and has published articles in *Middle Eastern Review of International Affairs, Turkish Studies Journal, New Perspectives on Turkey*, and *Middle Eastern Studies*, and is co-editor of *Re-Envisioning Egypt, 1919–1952 (2005)*.

Nandini Sundar received her Ph.D. from Columbia University, is a Professor at the Department of Sociology, Delhi School of Economics, and has published *Subalterns and Sovereigns* (1997), *Branching Out: JFM in India* (2001), as well as numerous articles.

William Westerman received his Ph.D. from the University of Pennsylvania, is the coordinator for outreach to immigrant communities and artists at the New York Foundation for the Arts, and the founder and creative director of Art Knows No Borders, Inc.

Shaking the Foundations of Education

An Introduction to Revolution and Pedagogy

E. Thomas Ewing

In 1925, at the age of seven, a young boy named Rolihlahla began attending a Wesleyan missionary school located not far from the eastern coast of South Africa. The first in his family to attend a formal school, this child was enrolled by his father, who "had the great respect for education that is often present in those who are uneducated." In preparation for his first day, Rolihlahla acquired new clothes: in place of the customary blanket, his father took a pair of his own trousers, cut off the legs, and tied them around the boy's waist with a string. According to the boy's later recollections, "I must have been a comical sight, but I have never owned a suit I was prouder to wear than my father's cut-off pants." In addition to the change of clothing, however, entry into the Western school also brought a new name. As was becoming customary in South Africa under colonial rule, indigenous names were not used by Whites, "who were either unable or unwilling to pronounce an African name, and considered it uncivilized to have one." On the first day, the teacher, Miss Mdingane, gave this child a new first name, which would, over the course of the next three-quarters of a century, become famous throughout the world: Nelson Mandela (Mandela 1995, pp. 6, 13–14).

Mandela's education continued through boarding schools for elite African children, higher education at the University College of Fort Hare, correspondence study for a law degree, and 18 years imprisonment on Robbens Island, which Mandela later referred to as "the University because of what we learned from each other" (Mandela 1995, p. 467). Mandela developed a deep appreciation for education, not only in terms of his own life, but also for its implications for all South Africans struggling for freedom, democracy, equality, and opportunity. In his memoirs, Mandela describes how his perspective was shaped by the school boycott of 1955, which challenged the institutionalization of white domination through curriculum and policies:

> Education is the great engine of personal development. It is through education that the daughter of a peasant can become a doctor, that the son of a mineworker can become the head of the mine, that a child of farmworkers can become the president of a great nation. It is what we make out of what we have, not what we are given, that separates one person from another. (Mandela 1995, p. 166)

While the boycott failed, in the sense that most African parents resigned themselves to apartheid schools, the government was forced to make some modifications in its policies and even Prime Minister Hendrick Verwoerd had to admit that, in Mandela's words, "education should be the same for all." Ultimately, Mandela argued, apartheid in education "came back to haunt the government in unforeseen ways," because these racially separated schools "produced in the 1970s the angriest, most rebellious generation of black youth the country had ever seen" (Mandela 1995, pp. 169–170, 483–484).

As these statements suggest, South African education could serve both liberating and repressive functions (Comaroff 1996, pp. 28–29). Educated in white-dominated schools integrated into the emerging apartheid system, Mandela recognized that his schooling, among other influences, provided the knowledge, integrity, determination, and confidence to succeed as a lawyer and a political leader, while also providing clear evidence of how African civilization and Africans as people were treated as inferior. In the movement against apartheid, schools provided both the instruments to be used in this struggle and the institutions, values, and structures against which this struggle was being waged.

Mandela's education led him to follow a revolutionary path. Committed at a young age to the cause of liberating South Africa from repressive rule by the white minority, Mandela became a leader in the African National Congress, and then the main proponent of the decision to renounce exclusively nonviolent tactics against the intransigent and oppressive regime. The discovery of plans for armed struggle led to his arrest and trial in 1964, which was followed by 27 years of imprisonment. While in prison, Mandela welcomed news of guerilla warfare as evidence that the African National Congress had "engaged the enemy in combat on their own terms" (Mandela 1995, p. 439). In 1976, Mandela responded to news of black schoolchildren's resistance with this statement smuggled from prison: "Between the anvil of united mass action and the hammer of armed struggle we shall crush apartheid and white minority racist rule!" (cited in Sampson 1999, p. 271). Even after his release in 1989, Mandela steadfastly refused to renounce violent tactics while the South African regime continued its repressive policies, insisting that it was "the reality and the threat of the armed struggle that had brought the government to the verge of negotiations" (Mandela 1995, p. 568).

Yet, within South Africa and throughout the world, Mandela has now become a powerful symbol of how political revolution can be achieved through peaceful reconciliation. In his autobiography, Mandela described his evolving understanding of the dialectic of liberation and oppression:

> It was during these long and lonely years that my hunger for the freedom of my people became a hunger for the freedom of all people, white and black. I knew as well as I knew anything that the oppressor must be liberated just as surely as the oppressed . . . When I walked out of prison, that was my mission, to liberate the oppressed and the oppressor both . . . For to be free is not merely to cast off one's chains, but to live in a way that respects and enhances the freedom of others. (Mandela 1995, pp. 624–625)

On this "long walk to freedom," Mandela has repeatedly identified education as a key to revolutionary change. In 1990, during a triumphal visit to the United States just one year after his release from prison, Mandela declared: "Education is the most powerful weapon you can use to change the world" (quoted in Howe and Lewis 1990, p. 1). In 1997, as the first democratically elected president of South Africa, Mandela declared that "the power of education" is essential to the processes of "nation-building and reconciliation":

> Our previous system emphasized the physical and other differences of South Africans with devastating effects. We are steadily but surely introducing education that enables our children to exploit their similarities and common goals, while appreciating the strength in their diversity. (Mandela 1997)

More recently, during a celebration of his eighty-fifth birthday, Mandela echoed his declaration from a decade ago when he made this pledge: "I will spend the rest of my days trying to help secure a more educated and healthier South Africa. Education is the most important solution that we can use to change the world" ("Mandela" 2003).

Embedded in these statements about education, as more generally in Mandela's life and in any struggle for meaningful change, is a crucial tension between pedagogy and revolution. The path through education into public activism, the emergence of schools as contested sites for domination and resistance, and the promise of new leaders to provide education for all are common themes across revolutionary cultures and pedagogical contexts (Comaroff 1996). Mandela's education as a revolutionary also demonstrates the tensions inherent in this relationship. Pedagogy can be both conservative and radical, just as revolutions can be liberating and repressive. Both categories are invested with complex and contradictory meanings; their juxtaposition offers important insights into their far-reaching reverberations.

This collection explores the tensions between and within processes of revolutionary and pedagogical change and continuity. By focusing on those enacting pedagogical activities in revolutionary contexts or pursuing revolutionary agendas in pedagogical contexts, these eight chapters provide an innovative and sophisticated exploration of complex directions and forces. These revolutions include the struggle for independence in the Philippines, the Russian revolution that led to the communist Soviet Union, the Egyptian campaigns against British colonial authority, the development of Kurdish national identity in the context of Turkey's modernization, radical and reformist educational movements in Western Europe and the Americas, the Palestinian struggle for self-determination, and the contemporary debate over national and religious identity in India. The subjects of analysis include "conventional" topics such as school policies and curricular content, as well as more "nontraditional" pedagogies such as public celebrations of holidays, participation in international exchange programs, and the incarceration of political activists. The interdisciplinary and transnational perspectives emerge from the explicitly comparative approach of each chapter, from the application of a wide range of disciplinary approaches, and from authors' locations that transcend narrow geographical or academic categories.

In this sense, the materials and interpretations presented in this collection truly "shake" the foundations of education, for they call attention to the embedded contradictions and tensions in the pedagogical project as well as the emancipatory and constraining implications of revolutions. As discussed in the final section of this introduction, these chapters explore, challenge, and certainly complicate the powerful assumption in the field of educational foundations that schools are inherently conservative institutions that, seemingly inevitably, reinforce the dominant structures of a given social order. By exploring revolutionary pedagogies and pedagogical revolutions from these interdisciplinary and transnational perspectives, this collection suggests new ways of considering the foundations of education in the worlds of the past, present, and future.

REVOLUTION

The revolutions examined in these chapters include social forces and political changes that transform structures, anticolonial movements that challenge external authorities by asserting national interests, state-directed transformations that impose "modernization" on "traditional" societies, and radical movements within ruling groups that pursue new directions of change from inside the dominant order. The category of revolution encompasses all these patterns because the category itself encompasses multiple meanings, processes, and outcomes. The Russian revolution in 1917, the establishment of Filipino independence, or the modernization of Turkey are large-scale transformations; the decision to educate girls, the act of writing by a political captive, learning a traditional approach to drumming, or putting a uniform on a teacher are small-scale changes. Yet, they are all revolutionary, because they each involve a transformation—or a reproduction in a different context—of an established or imposed order. To appreciate the complexity of this category, this section provides the contextual background needed to understand the case studies while also highlighting distinct features of each revolution.

In the case of the Russian revolution (chapter 2), E. Thomas Ewing focuses on two texts produced on either side of two revolutionary divides: Moscow schoolteacher E. Kirpichnikova's article preceded the 1917 revolution that ushered in a fundamentally new political system, while educator

A. Savich's 1939 article appeared a decade after Joseph Stalin's "revolution from above" resulted in even more far-reaching changes in Soviet society and culture. The comparison of these two articles reflects, on the one hand, the continuity of discourse that transcended revolutionary transformations, as both authors dealt with the similar question of how schools could pursue equity while recognizing apparent differences between boys and girls. On the other hand, such a comparison also illustrates how the major changes of the Russian revolution found expression on the more subtle level of "taken-for-granted" assumptions. Whereas Savich wrote with confidence that coeducation was necessary because it was consistent with the goals of socialism, Kirpichnikova's article was part of a real debate about coeducation in the face of entrenched conservative opposition and government reluctance to pursue such a seemingly revolutionary change. While neither the 1917 revolution nor Stalin's "great break" had gender equity as an objective, these articles demonstrate how revolutionary changes in political and ideological spheres had indirect, yet equally significant, repercussions in the pedagogical realm. Even here, revolution and pedagogy maintained their contradictory relationship, as a dramatic reversal in public policy—from actively opposing to strongly supporting coeducation—stood in sharp contrast with persistent similarities in pupil behavior and teacher practices.

Cati Coe (chapter 4) also takes up the challenge of interpreting institutional revolutions. In Ghana, the impetus for educational change came in part from a change in the regime, even as the intentions and implications of the new leadership were constrained by the extent of compliance, coordination, or resistance at the local level. This revolution began with the seizure of power in 1981 by the Provisional National Defence Council (PNDC), which replaced a succession of military governments that had governed Ghana for most of the two decades following independence from British colonial rule. During the period covered in this chapter, the PNDC combined anticolonial and anti-Western rhetoric with neoliberal accommodations to the demands of global capital and deliberate efforts to extend government power domestically. Ghana thus provides a case study of an institutional revolution, in which schools become instruments intended to reinforce, rather than challenge, the interests of an elite. As this chapter also demonstrates, however, the effort to use cultural policies to achieve definite aims remained dependent on existing structures, relations, and interactions. While the Ghanaian government may have seen culture as a sphere independent of global economic forces, and thus more susceptible to control by a government searching for markers of authenticity, Coe's interviews with teachers, observations of classroom rituals, and close reading of texts demonstrate the persistent contradictions of this revolutionary context.

Twentieth-century Turkey provides a different model of institutional revolution. As chapter 6 by Yucel Demirer clearly demonstrates, however, the process of change was neither unilinear nor all-inclusive. In the new Turkish republic that emerged following the Ottoman defeat in World War I, a series of state-initiated reforms pushed a radical "Westernizing" agenda in such spheres as religion, women's rights, and education. The ideology of

"Kemalism" that emerged in the 1930s and persisted for decades to follow combined populist nationalism with state-directed economic modernization, cultural secularization, and social transformation. In order to promote a new Turkish national identity, however, the government deliberately, and at times forcibly, suppressed alternative identities perceived as obstacles to this modernizing agenda. The tensions of the revolutionary project thus found expression in the contradictory location of the Kurdish minority, for whom the suppression of rights and the denial of opportunity served as a stimulus for strengthened national identity and broader dissatisfaction with the state-building project. As Demirer indicates, the articulation of national identities through a public discourse of celebrations illustrates the dilemmas of an institutional revolution, which acquires authority only to the extent it persuades the population of the legitimacy of promised transformations.

In his discussion of popular pedagogies as forms of political engagement, William Westerman discusses societies ranging from the avowedly revolutionary, such as Cuba in the early 1960s and Nicaragua in the early 1980s, through societies undergoing dramatic social changes, such as nineteenth-century Denmark and the southern United States in the 1960s, to those governments that forcibly suppressed radical pedagogies, such as Brazil in the 1960s and 1970s. Westerman's study begins with a Danish movement to define a national identity in opposition to its dominant neighbors and consistent with democratic, egalitarian, and popular ideals. From this perspective, the shaping of a modern nation drew upon and also reinforced the spread of popular schools, which taught children and adults to think of themselves as part of this larger community. A century later, revolutionary governments such as Cuba's Castro regime or the Sandinistas in Nicaragua drew upon this tradition to make popular education into a means of mobilizing the masses, especially the rural peasants, to support the radical policies of their new government. These regimes sponsored literacy campaigns in which the teaching of basic skills were accompanied by political indoctrination, yet, as Westerman argues, the structure of the lessons mattered as much as the content of the materials. Popular pedagogies also emerged, however, in less conducive environments, including the segregated regime of the southern United States and the military dictatorships of Latin America. In these contexts, where state-supported political and social structures maintained repressive regimes, alternative pedagogies such as the Highlander School in the United States and the "pedagogy of the oppressed" in Latin America attracted followers who saw popular education as a path to political empowerment. While none of these states ever fully realized the principles of the folk school and popular pedagogy movements, the lessons learned by followers became part of the broader cultural, educational, and political context in which these individuals sought to create a better society.

The Filipino struggle for national independence, first against the Spanish and then against the United States, defines the revolutionary context for Roland Sintos Coloma's study of the life and work of Camilo Osias (chapter 1). The guerilla war waged against Spanish imperialism, which ended in 1898 with annexation by the United States, promoted a growing sense of Filipino nationalism emphasizing a distinct sense of identity. While United States authorities encouraged certain kinds of national development, they also preserved and even strengthened the existing colonial relationship between indigenous communities and imperial power. The contradictions of these policies were evident in the life of Osias, whose education and career were promoted by specific American policies, such as English-language schools, the *pensionado* program, which sent Filipino students to American colleges, and the creation of an indigenous corps of educators and school administrators. Yet even as Osias and many other Filipinos pursued these opportunities, a distinct national identity emerged that challenged this unequal relationship. The Filipino struggle for national independence was thus a revolution in which indigenous elites drew upon—or "identified with," in Coloma's theoretical framework—even as they challenged—or "disidentified with"—this relationship between subordinate actors and dominant forces. The emerging identity of Osias thus provides a means by which to understand the strategies of a subordinate group rebelling against a dominant system with which they share key values and more importantly seek to emulate. Philippine independence in 1946 marked the end of the anti-imperial struggle, yet Filipino national identity continued to be shaped by the legacies of this colonial relationship.

The Egyptian struggle for national independence followed a similar trajectory, as advocates of self-determination challenged English authority even as they sought to build upon certain aspects of this colonial relationship. In chapter 3 by Barak A. Salmoni, multiple revolutionary processes occur simultaneously: the independence campaign begun in 1918 by national leaders known as the *Wafd* (delegation), a cultural revolution intended to overcome "backward" indigenous traditions in pursuit of modernity, and a feminist movement that associated women's emancipation at home and in society with progress toward national self-determination. In Egypt, as in the case studies discussed in the other chapters, internal as well as external influences shaped revolutionary trajectories. Prominent Egyptians, like nationalist leaders in India, southern Africa, and the Arab world, hoped that the Versailles peace conference would fulfill wartime promises of national self-determination in the colonial sphere. The disillusionment with the European response served to stimulate a nationalist movement that would achieve independence in 1922 and continue to pursue full emancipation from English control over the next three decades. By viewing European power as both the inspiration for and the obstacle to this campaign for national self-determination, Egyptian leaders occupied a contradictory position relative to the outside world. In Salmoni's discussion of texts by educators such as Huda al-Sha'rawi, Ahmad Lutfi al-Sayyid, Amir Boktor, and Asma' Fahmi, the contradictions of this revolutionary position were evident in the expectation that a genuine Egyptian woman would be educated in a Western-style school with an Egyptian, Arabic, and Islamic curriculum that would prepare her to serve the nation by becoming an ideal daughter, wife, and mother. Yet, these contradictions should not obscure the significance of these revolutionary appeals for the education of girls and the equality of women in the context of the developing Egyptian national community.

The most contemporary revolutions explored in this collection, involving present-day India and Palestine, share the characteristic of being directed against governments that were themselves established through anticolonial struggles. In his study of Palestinian political captives in Israeli jails, Esmail Nashif examines the period after the first *intifada* (uprising) of the 1980s, when an older "generation" of Palestinian activists imprisoned in the late 1960s and throughout the 1970s were joined by a new wave of younger activists. The Palestinian captive community has broad political objectives—to establish an independent state—but also has more subtle transformative goals—including a redefinition of morality and interpersonal relations within the Palestinian movement, an articulation of alternative identities, and, most importantly for the purposes of this collection, a strategic use of education to resist, evade, or mediate the oppressive power and authority of the prison. Yet the campaigns examined by Nashif work in contradictory ways as well. To the extent that the prison itself becomes a site for transformative practices, both the structures of imprisonment and the agency of captives become forces for change. Revolution thus acquires dual meanings: a political campaign waged by a nationalist movement against an occupying force and a transformative project in which institutional forces as well as personal agency contribute to the formation of alternate subjectivities. By exploring the experiences and language of these political captives, Nashif explores the tensions as well as the potential of revolutions that occur simultaneously on the multiple levels of individual identity, institutional power, and political mobilization.

The Indian case study by Nandini Sundar asks how a state with revolutionary origins has been transformed from within by advocates of a more exclusive vision of political community. Until its removal by the voters in the 2004 elections, India's government led by the Bharatiya Janata Party (BJP) pursued a Hindu nationalist platform that has departed from the more inclusive and intercommunal approach pioneered by Mahatma Gandhi under British colonialism and sustained to a great extent subsequently by the Indian National Congress. During the half-century after independence, however, alternative political movements, including the Rashtriya Swayamsevak Sangh (RSS), have advocated extreme versions of Hindu identity politics. In chapter 8 by Sundar, the efforts of the RSS to transform Indian education through such instruments as the history curriculum, the creation of alternative schools, and the celebration of sectarian holidays constitute a kind of revolution from within, as a political organization pressures the government to pursue a radical new policy. Sundar's study demonstrates how the underlying forces that shaped the Russian, Turkish, and Ghanaian revolutions and the struggles for national self-determination by Filipinos, Egyptians, Palestinians, and Kurds remain relevant in the contemporary world. The shift to identity politics, the emergence of nongovernmental organizations, and the structures of globalization have not eliminated fundamental questions of who holds power, whose interests are being served, and what people are willing—or unwilling—to do in order to pursue—or obstruct—significant changes.

Pedagogy

Pedagogy incorporates pupil interactions and teachers' practices, textbooks and curricula, celebrations and rituals, and the discourse of educational administrators and policy advocates. Pedagogy thus includes, but also transcends, the classroom and the school. One of the shared goals of these chapters is to follow Christine Heward's lead in moving beyond "structural" issues, such as policy and enrollment, to examine the "social relations" embodied in the pedagogies that occur in revolutionary contexts (Heward 1999, p. 3). In each chapter, therefore, pedagogy is ultimately a matter of relationships: between political captives and prison structures, in chapter 7 by Nashif; between the advocates and opponents of an equal education for girls, in chapters 2 and 3 by Ewing and Salmoni; between the forces promoting a dominant culture and those engaged in various forms of resistance, in chapters 1, 5, and 6 by Coloma, Westerman, and Demirer; or among proponents of different strategies of cultural transmission, in chapters 4 and 8 by Coe and Sundar.

Yet even this outline simplifies relations that are—both in the unique circumstances studied in each chapter and in the collection taken as a whole—illustrative of the layers of complexity inherent in the category of pedagogy. In chapter 4, the focus is on different approaches to teaching and learning culture in contemporary Ghana. Educational reforms implemented by the state made "cultural studies" into a required part of the school curriculum. In this context, however, the effort to use schools to disseminate a definition of culture that legitimated the PNDC and its neoliberal policies conflicted with the multiple definitions of culture articulated and propagated within communities by chiefs, teachers, and other elders. The complex relationship between central state institutions and local authority ensured that schools became sites for struggles, however concealed or denied, over the meaning of "authentic" culture. Approaching schools as spaces in which competing meanings were articulated, enforced, and contested by multiple actors, chapter 4 reveals the complexities of pedagogy in a revolutionary context. By comparing the formal curriculum with behaviors observed in a classroom and with teachers' own explanations of why they acted and spoke in certain ways, Coe deconstructs the pedagogy of cultural studies into its multiple, and to some extent contradictory, elements.

In Nashif's study of Palestinian political captives, the pedagogical context is radically different. Prisons censored information, restricted communication, and made totalizing claims on space, bodies, and behavior. Yet the prison itself became, in the words of former captive Hasan Abdallah, a site for learning that was "far more sophisticated" than the university. In this context, captives developed an approach known as *thaqafah* (culture) to reconstitute, reaffirm, and strengthen Palestinian identity. Reading/writing were located at the center of *thaqafah* as practices that simultaneously refuted the imposed identity of "captive" and asserted the alternative identity of Palestinian national activist. Pedagogy in this context thus refers to a complex set of informal mechanisms of teaching and learning: older captives teaching new arrivals about survival techniques, captives who spoke foreign languages teaching them to others, lessons in military science that drew on the texts of revolutionary commanders, exposure to liberation texts produced in other contexts, and organized efforts to propagate the distinct ideologies of different Palestinian organizations.

Folk schools and popular pedagogies emerged as efforts to transform societies from below, through processes of teaching others to have faith in their own capacities. In each case examined by Westerman, leaders articulated theories that attracted followers into institutions, movements, and organizations dedicated to radical change: N. F. S. Grundtvig's Folk Schools designed to integrate Danish peasants into a sense of national community, Myles Horton's Southern Mountains School, which taught a generation of American labor activists and civil rights leaders to challenge entrenched structures of political and economic oppression, and Paolo Freire's method of teaching the oppressed to read the word as well as the world. In each case, pedagogy was simultaneously grounded in an immediate social reality while also seeking to transcend and transform the confines of the surrounding context. As Westerman argues, the movement of pedagogies from theory to practice involved both radical challenges to the status quo and gradual accommodations to reconstituted forms of oppression. Yet, the theories themselves mattered, regardless of the outcome, because they placed everyday practices and language at the center of the educational process. In this sense, the suppression of an organization, the end of a school, or the cooptation of a movement become part of a continuing dialogue, as the inspirational principles as well as the cautionary lessons reemerged in other contexts to reaffirm the dialectical relationship between pedagogy and revolution.

Pedagogy also refers to the efforts of "nongovernmental organizations" influence the content and context of education. In contemporary India, a network of private schools, a growing presence on educational boards, and direct control of public organizations allows the RSS to promote an explicitly anti-Moslem and anti-Christian agenda that defines a vision of history and community exclusively Hindu in orientation and composition. Through an exploration of contemporary educational discourse as well as field studies in the Chhattisgarh state, Sundar describes how the RSS and its front organizations disseminate their message of religious supremacy, cultural intolerance, and militant identity. Observations of classroom practices and school rituals, conversations with teachers and parents, and textual analysis of educational materials illustrate Sundar's argument that the RSS is pursuing a revolutionary transformation of and through the schools. The fact that RSS schools are considered temples—a designation articulated and reinforced by behavior, dress, and speech—confirms the extent to which this organization challenges both traditional and modern definitions of public education. Sundar's conclusion offers a more pessimistic evaluation of how revolutionary and pedagogical strategies serve to affirm, even as they conceal, the powerful operations of a hegemonic system.

The broadest definition of pedagogy emerges in the study by Demirer, who focuses on the ways in which elites use cultural traditions to reinforce bonds of loyalty and obedience. In the Newroz/Nevruz celebration, which marks the coming of spring, the shared objective of constituting a common culture is disrupted by sharp divisions over the meaning, character, and purpose of these celebrations. Demirer interprets the Kurdish celebration, Newroz, as a pedagogical and political instrument that challenges the hegemony of the state in the cultural realm. While the celebration has many elements, the emphasis on articulating a Kurdish identity through past traditions, current customs, and future visions constitutes an alternative forum for civic education. In particular, Demirer argues, the Newroz celebrations allow an older generation of Kurds to articulate a separate, and at times revolutionary, vision of national identity denied to the younger generation by state institutions, most particularly the schools. By contrast, the Turkish celebration, Nevruz, has become an extension of the official pedagogy as disseminated through the media, by politicians, and in the schools. In this case, a "pedagogy of domestication" suppresses the articulation of a Kurdish identity by appropriating this holiday and reconstituting it as a source of legitimacy for the Turkish political order. For both the Kurdish community and the Turkish state, therefore, celebrations serve important pedagogical and revolutionary functions, even as the content, objectives, and implications depart in radically different directions.

Pedagogy also occurs in the complex relationship between descriptions of school practices and normative statements of what should be happening in schools. In Russian/Soviet and Egyptian discourse on gender and education, for example, observation and advocacy both assumed pedagogical functions. In chapter 2, an experienced teacher and a school inspector described behaviors that supported their own interpretations of what should, or should not, be happening in classrooms, on playing fields, or among pupils. In this sense, pedagogy occurred on two distinct, yet related, levels: what happens in classrooms, and the discursive representation of what should, could, or might happen. In a similar manner, Salmoni's analysis of texts advocating the education of Egyptian girls offers a broad perspective on how this seemingly narrow question became inseparable from the challenge of ensuring that education best served the emerging national interest. These materials were both descriptive and prescriptive, as each author's arguments for a certain kind of girls' education included some assertion of how implementing these changes would enable schools to promote national interests. In both chapters, educational discourse itself assumes pedagogical functions through these efforts to disseminate, replicate, and transform the context and content of schooling.

For the Filipino educator Osias, pedagogy and revolution were woven into an entire life narrative. Osias borrowed important concepts and language from American educational discourse even as he sought to overcome their exclusionary thrust by defining a more empowering identity for teachers and pupils. By constructing Filipino pedagogy in reference to, but also in opposition to, the hegemonic discourse he experienced as a pupil in colonial schools, as an exchange student in an American university, and as an educational superintendent in the final years of United States occupation, Osias deconstructed the dominant structures of power and knowledge in order to reconstruct an alternative identity. Pedagogy in chapter 1, thus, refers not only to multiple levels of schooling and to such formal devices as the curriculum, textbooks, and teacher training programs, but more broadly to the ways in which learning shapes identities on both individual and national levels. The man called "the Patrick Henry of the Philippines" by an American classmate almost forty years before independence thus embodied the complex relations between pedagogy and revolution that are recurring themes in all these chapters.

INTERDISCIPLINARY AND TRANSNATIONAL PERSPECTIVES

Understanding revolution and pedagogy requires interdisciplinary and transnational approaches that lead to a reconsideration of the social foundations of education. The authors draw upon a variety of disciplinary perspectives—education, history, anthropology, gender studies, political science, and folklore—each of which position subjects and structures at the center of analysis. While focusing on specific geographical regions, chronological periods, and thematic subjects, these chapters all address a broader set of issues that transcend these categories of space, time, and content. The interdisciplinary and transnational perspectives thus inform and are located in each chapter, while the accumulated effect of the collection offers new insights into the relationship between revolution and pedagogy.

Chapters 1–3 and 5 address the relationship between revolution and pedagogy from a historical perspective, where the key sources are documents written by educators. These chapters trace the dynamic of change over time: the evolving discourse of educational and national liberation in Egypt, the debate on coeducation in revolutionary Russia, pedagogical theories promoted by educator-activists in Western Europe and the Americas, and the campaign for political independence in the Philippines. Historical analysis encourages the close reading of texts that recognizes layers of meaning while also acknowledging the complex interplay of structure and agency over time. In addition, each author explores issues that remain significant in the present: the relationship between personal and national liberation, the gendering of discourses of autonomy and dependency, and the persistence of economic dependence and cultural inequity in contexts of political self-determination.

Coe and Demirer draw on research methods of participant observation, supplemented by source analysis of historical materials and a review of scholarship on national, educational, and cultural developments. Based on ethnographic research in classrooms in Akuapem, Coe interprets the teaching of culture through such mechanisms as recreating traditional rituals, retelling community tales, and reproducing cultural practices. Situating her study in a region where she conducted extensive fieldwork, Coe uses this base of empirical evidence to address the broader question of the multiple and contested meanings of culture in African classrooms and communities in a globalized world context. Demirer combines a historical analysis of the evolution of

Newroz/Nevruz celebrations with first-hand observations of these celebrations as they are practiced in contemporary Turkey. The latter method is especially fruitful, as chapter 6 provides a clear sense of the emotional and social meanings that inform and embody the political, cultural, and historical significance of these celebrations. These two chapters thus demonstrate that revolution and pedagogy need to be interpreted as lived experiences, where what is seen, done, and said provide insights into the meanings constructed by participants and subsequently by interpreters.

Chapters 7 and 8 involve both an analytical reading of texts produced by revolutionary participants and direct engagement with political movements of the present day. In interviews, essays, and fiction, Palestinian captives describe their experience in prison and embody the transformative implications of these simultaneous processes of pedagogy and revolution. By interviewing subjects and reading texts, Nashif participates in a form of *thaqafah* by making the act of reading/writing central to the analytical and interpretive process. In this sense, the pedagogical act that begins in the prison continues into this chapter, which challenges readers to engage in the same process of constituting subjects in a context defined by unequal institutional power dynamics. In a similar manner, Sundar approaches the topic of RSS schools from the perspective of an academic and educator committed to a vision of Indian schooling that is democratic, inclusive, and tolerant. By examining the language used to recruit pupils, through analysis of curricular content, and by comparisons with extremist movements in other contexts, Sundar effectively demonstrates how the educational pronouncements of the RSS construct, while also seeking to conceal, an ambitious political agenda with far-reaching implications not only for India but also for the rest of the world.

The transnational character of these chapters is best revealed by a reading that compares approaches and contrasts interpretations. The concept of *transnationality* allows for examination of specific topics within a *national context* while also recognizing processes, patterns, and categories that *transcend* the boundaries of the nation-state (Thelen 1999, p. 968). All of these revolutions were unique, yet each occurred in a context shaped by the implications of earlier transformations and the potential of future developments. The Russian revolution, the Filipino struggle for independence, Danish folk schools, and the Egyptian nationalist movement were significant parts of a broader movement in the late nineteenth and early twentieth century toward national self-determination. Neoliberalist cultural policies in Ghana, tensions between the Kurdish community and the Turkish state, state-supported literacy campaigns in Cuba and Nicaragua, the civil rights movement in the southern United States, the resurgent Hindu nationalism in India, and the Palestinian struggle for self-determination illustrate how conflicts over identity within nation-states are also part of global discourses of identity-politics. While each revolution occurred within a distinct region, community, and culture, they all involve processes of domination and the possibilities of liberation.

Pedagogy is an excellent site for exploring these transnational and interdisciplinary perspectives, because schools as institutions and learning as a process assumes a certain universality, even as the actual content, the specific experiences, and especially the broader implications are made meaningful by context, agency, and practice. Thus debates on women's education in Egypt and Russia/Soviet Union, movements for national self-determination in the Philippines, struggles for autonomy by Kurds and Palestinians, self-education schools and circles in Western Europe and the Americas, and state-directed cultural programs in India and Ghana share certain important characteristics, yet also have their own unique trajectories due to political context, cultural content, and global position. This book's approach also reflects the positions of the authors, each of whom has crossed geographical as well as disciplinary boundaries to research and write about revolution and pedagogy. The authors' interest in transnational and interdisciplinary perspectives is thus more than academic, and embodies instead a broader concern about local and global interrelationships in the past, present, and future. These perspectives thus allow for an engagement with transcendent processes as well as understanding of the particular shape and specific implications of each case study.

Shaking the Social Foundations of Education

A common theme in the literature on the social foundations of education is the fundamentally conservative nature of schools. In 1932, for example, George Counts, one of the first American scholars to define social foundations of education as a distinct field, declared emphatically: "Almost everywhere [the school] is in the grip of conservative forces and is serving the cause of perpetuating ideas and institutions suited to an age that is gone" (Counts 1969 [1932], pp. 4–5). A similar point is emphasized in more recent studies of American schools and society, as in this statement by Daniel Selakovich: "Schools are not revolutionary agents of change but tend to support the political, religious, and economic values which exist in the society" (Selakovich 1984, p. 141). Textbooks on educational foundations thus share a general consensus that the prevailing pattern in most societies is for schools to reproduce the dominant order, both through the creation of productive forces (i.e., working-class jobs for lower-class children, executive and professional positions for upper-class children) and the perpetuation of an ideology of social stratification, political hierarchy, and cultural hegemony. This process of reproduction may be framed in a liberal rhetoric of individual rights, creation of opportunities, and respect of multiculturalism, or may be framed in a conservative rhetoric of developing a strong national identity, denying alternative visions that undermine unity, and asserting the need for order, stability, and discipline.

Even authors of textbooks designed for the mass market of state-mandated teacher education courses, as in the following statement, view the school as a fundamentally conservative institution:

> Schooling plays an important role in teaching and legitimating a society's ideology. The ideology served by the public school is almost inevitably the dominant ideology of the larger society . . . Schooling prepares people to

participate in a society's political economy and share its dominant ideology, but by doing so, it may further disadvantage those from the less-advantaged groups while contributing to the already privileged position of the more powerful. (Tozer et al. 2002, p. 10)

Of course, some scholars and educators promote alternative practices that question or subvert the dominant order (see Westerman, chapter 5, this volume, as well as Shor 1993; Giroux 1981; McLaren 1989). Quite often, however, even these advocates of emancipatory and transformative possibilities concede that "most" schools tend to reproduce the dominant values of the surrounding community and the larger social structures. (See review of theories in Izquierdo and Minguez 2003.)

These chapters offer a different perspective on this crucial subject. This book does not argue that pedagogy is, could, or even should be revolutionary. Rather, this book explores the complex possibilities of education in revolutionary contexts and of revolutions in educational contexts. In each chapter, an educational foundation has been "shaken" by broader social, political, and international forces: a communist revolution in Russia, campaigns for national independence in the Philippines, Denmark, and Egypt, state-directed modernization in Turkey, struggles for self-determination among Palestinians, African Americans, and Kurds, and responses to a new world order in Central America and Ghana. Educational foundations are also "shaken" by new pedagogies: the pursuit of female education as an aspect of modernization, the embrace of a pedagogy of national identity, the public rituals of community celebrations, literacy instruction as empowerment, the lessons learned in captivity, and contests over the meaning of cultural identities. The argument made by these chapters separately and as a whole is that the meanings of pedagogy and revolution can be better understood by examining their intersections and interactions. Rather than accepting that schools serve primarily as structures of reproduction, or identifying instances of resistance in ways that confirm, however indirectly, this larger claim, these chapters make the revolutionary implications of pedagogy and the pedagogical implications of revolutions into questions for analysis.

This introduction began with analysis of how Nelson Mandela synthesized within his life struggle a revolutionary commitment to change through a process of teaching and learning. Yet, Mandela also illustrates the limits of this synthesis in the emerging world of the twenty-first century. Mandela's own statement cited above, "Education is the most powerful weapon you can use to change the world," brilliantly captures the revolutionary potential of pedagogy. Yet, this slogan, and with it the underlying message, is in imminent danger of being reduced to a cliché. Schools, organizations, and corporations have adopted this quotation as a mission statement, but with little evidence of the revolutionary commitment and transformative pedagogy expressed through Mandela's life.

For individuals, communities, and societies in a seemingly "post-revolutionary" age, the temptations and expectations of acceptance and conformity are extremely powerful. Yet, this volume offers radically different perspectives. This book is not a call to revolution; if anything, these chapters suggest that the historical experience of revolution is repressive as well as liberating, stultifying as well as emancipating, and destructive as well as constructive. Pedagogy embodies these same contradictions, yet also provides an alternative source of visions, practices, and interventions (Comaroff 1996). Educational foundations have been and will continue to be shaken by revolutionary and pedagogical forces, with implications that remain as far-reaching as they are unpredictable. In an era when education is increasingly global in content and reach, this book represents a timely intervention by adopting transnational perspectives on pedagogical and revolutionary dynamics.

This book should also be read with an awareness of the extraordinary vision of the approximately 1.5 billion school age children and youth in the world today. At a time when the percentage of young people is growing to unprecedented levels on a global scale, the relationship between pedagogy and revolution promises to become ever more central in the coming decades. As the inheritors of the revolutionary and pedagogical currents explored in this book and as citizens of an increasingly transnational world, these children will become the educated and the educators of the future. The contributors see their research not only as a scholarly effort, but also as a commitment to a new vision of this global community. In this respect, this book is well situated to guide, inform, and inspire readers to think about revolution and pedagogy from perspectives that are themselves both pedagogical and revolutionary.

References

Comaroff, Jean. 1996. "Reading, Rioting, and Arithmetic: The Impact of Mission Education on Black Consciousness in South Africa." *Bulletin of the Institute of Ethnology. Academic Sinica* no. 82, Autumn, pp. 19–63.

Counts, George S. 1969. *Dare the School Build a New Social Order?* New York: Arno Press. Reprint of 1932 edition.

Giroux, Henry A. 1981. *Ideology, Culture, and the Process of Schooling.* Philadelphia: Temple University Press.

Heward, Christine. 1999. "Introduction: The New Discourses of Gender, Education, and Development." In *Gender, Education, and Development. Beyond Access to Empowerment.* Heward and S. Bunwaree, eds. London: Zed Books, pp. 1–14.

Howe, Peter J. and Diane E. Lewis. 1990. "Mandela and Boston Embrace in a Daylong Celebration of Unity," *Boston Globe* June 24, p. 1.

Izquierdo, Honorio Martin and Almudena Moreno Minguez. 2003. "Sociological Theory of Education in the Dialectical Perspective." In *The International Handbook on the Sociology of Education. An International Assessment of New Research and Theory.* Carlos Alberto Torres and Ari Antikainen, eds. Lanham: Rowman and Littlefield Publishers, Inc., pp. 21–41.

Mandela, Nelson. 1995. *Long Walk to Freedom. The Autobiography of Nelson Mandela.* Boston: Little, Brown and Company.

———. 1997. "Speech by President Mandela at the Education Africa, Presidential and Premier Education Awards." South African Presidential Website: <www.polity.org.za> (accessed August 2003).

"Mandela to Continue fighting poverty." 2003. <Iafrica.com> (accessed August 2003).

McLaren, Peter. 1989. *Life in Schools. An Introduction to Critical Pedagogy in the Foundations of Education*. White Plains: Longman.

Sampson, Anthony. 1999. *Mandela. The Authorized Biography*. New York: Alfred A. Knopf.

Selakovich, Daniel. 1984. *Schooling in America. Social Foundations of Education*. New York: Longman.

Shor, Ira. 1993. "Education is Politics. Paulo Freire's Critical Pedagogy." In *Paulo Freire. A Critical Encounter*. Peter McClaren and Peter Leonard, eds. London: Routledge.

Thelen, David. 1999. "The Nation and Beyond: Transnational Perspectives on United States History," *The Journal of American History* vol. 86, no. 3, December, pp. 965–975.

Tozer, Steven E., Paul C. Violas, and Guy Senese. 2002. *School and Society. Historical and Contemporary Perspectives* 4th ed. New York: McGraw-Hill.

1

Disidentifying Nationalism

Camilo Osias and Filipino Education in the Early Twentieth Century

Roland Sintos Coloma

Representing the state of Illinois at a Midwest inter-normal oratorical competition on May 8, 1908, the 19-year-old Camilo Osias began his speech "The Aspiration of the Filipinos" with the phrase: "All nations and individuals love liberty and independence; they hate servitude and restraint" (Osias 1908). Ten years after the United States gained control of his native country at the end of the Spanish–American War, Osias addressed the American spectators not as a mere contest participant, but as a living symbol of intellect and civility, a testimony to the Filipino capacity for self-rule. He stressed the American ideals of freedom and equality to win the hearts of the audience and the votes of the judges as well as to appeal for his country's independence. Whereas Spain, the former imperialist ruler of the Philippines, became an emblem of cruelty and injustice, he pointed to America's commitment to liberty through its Declaration of Independence, its eradication of slavery, and its pledge of "Philippines for the Filipinos." The speech impressed and won over the judges who awarded him the grand prize. Demonstrating a sophisticated understanding of the tensions in the American colonial project and challenging the unequal relationship between the United States and the Philippines in the early twentieth century, Osias performed a type of nationalism, which appropriated and refashioned the codes and language of the dominant power in order to advocate for the sovereignty of his country.

This chapter focuses on the schooling and career of Camilo Osias (1889–1976) in order to explore the themes of revolution and pedagogy and, in particular, the intersection of education, imperialism and nationalism in the context of anticolonial revolution. Education is broadly defined as pedagogical engagements where teaching and learning occur both in the formal school settings and in the informal nonschool environments. Osias's pedagogical engagements, for instance, took place in a Spanish grammar school and an American-run high school, in government-sponsored matriculation at U.S. universities, in the Philippine educational system, and throughout the Filipino quest for self-determination. While this chapter does not portray Osias as a representative of all Filipinos during this time period, he serves as an example of how it was possible to utilize a Western education yet remain committed to the nationalist struggles of his home country. His successful navigation of the colonial tensions enabled him to emerge as a critical figure in the educational and political history of the Philippines. A leader of the "Filipinization" movement in the public education system, he became the first Filipino school division superintendent, the highest-ranking Filipino in the Bureau of Education as its assistant director, and the first Filipino school textbook author. Ultimately, Osias demonstrated how education was a contested site for anti-imperialist struggles that intersected with the more familiar patterns of political mobilization, sociocultural contestation, and subaltern strategies of resistance (Cabral 1994; Fanon 1963; Ileto 1979; McClintock 1995; Said 1993; Scott 1985).

Recognizing these multiple layers of anticolonial revolution draws attention to the space of "imperial encounters," which interconnects the West and its colonies—in this case, the United States and the Philippines—within a single analytical field. The concept of imperial encounters emphasizes the physical and figurative crossings and confrontations of metropolitan and colonial peoples, materials and cultures within and between their geopolitical boundaries. Drawing from the emerging postcolonial frameworks in the field of history (Cooper and Stoler 1997; Guha 1982; Jacobson 2000; Stoler 2001; Wolfe 1997) and consistent with the recent projects in Philippine and Filipina/o American Studies (Alidio 2001; Campomanes 1995; Choy 2003; Espiritu 2000; Fujita-Rony 2003; Rafael 2000; Salman 2001; San Juan 2000), this interpretive and methodological move produces "a history within and between," which disrupts conventional historiographies that have distinctly separated the histories of the metropole and the periphery and have rendered the effects of imperialism as solely unidirectional toward the colony. While postcolonial studies has generated significantly fresh understanding and analysis of transnational networks of power, oppression, and resistance in historical and current periods (Young 2001), it has only been since the early 1990s that a critical mass of academic work has focused on the operations of U.S. imperialism (Kaplan and Pease 1993; King 2000; Singh and Schmidt 2000).

The scholarly literature on the history of relations between the Philippines and the United States reinforces the notion that imperial encounters affected only the Philippines and not the United States, and reveals a dichotomy regarding the impact of American education in the archipelago. One side argues that the American common school system was an improvement over the private, elitist Spanish version and ushered in literacy and democracy in the country (Karnow 1989; May 1980). The other side contends that the American curriculum served as a "mis-education" since it depicted the United States as a benevolent hero that rescued the country from Spanish theocracy and native primitivism, yet concealed its ulterior military, economic, and political motives in Asia (Constantino 1966). This chapter participates in this historiographic debate and situates Filipino education under U.S. rule as neither Western/assimilationist nor indigenous/separatist. It shows, instead, how Camilo Osias navigated between these two competing forces and created a third space of possibility that utilized the dominant symbols and rhetoric of power in order to articulate and enact the politics of the oppressed and marginalized.

In interrogating the contested spaces of education in terms of the experiences and actions of an individual, this chapter suggests that the nationalist strategy for education and self-determination under colonial conditions was more about "disidentification" (Muñoz 1999) as opposed to the more conventional processes of identification or counter-identification. To become a revolutionary educator, Osias had to first learn, then distance himself, and finally use the knowledge acquired from Filipino peasants, Spanish clergy, American teachers, and the leaders of the Philippine sovereignty movements. The revolutionary implications of these various forms of pedagogy were realized only through the construction and performance of an identity that worked within and against the relational discourses and structures between U.S. colonialism and Filipino nationalism.

HISTORY AND NATIONALISM

Nationalism is a powerful revolutionary strategy against imperialist subjugation since "nations inspire love, and often profoundly self-sacrificing love" that ignites people's pride and courage to defend their countries and fight for their freedom (Anderson 1991, p. 14). Benedict Anderson's (1991) four typologies of nationalism—creole, vernacular, official, and last wave—offer a potentially relevant frame to interpret the history of the struggle for Philippine independence (Agoncillo 1990; Schirmer and Shalom 1987). The Philippine islands came under Spanish imperialist rule with the arrival of soldiers, merchants, and priests by the mid-1500s. By the late 1800s, European-educated Filipino elites began to clamor for political rights, particularly for representation in the Spanish metropolitan legislature. As members of the upper class from mixed *indio*/native, Spanish or Chinese backgrounds, they occupied a dual position in the colony; as economically and politically exploited and as stabilizers of imperialist control and status quo. Dr. Jose Rizal, the country's martyred hero, was a proponent of this *creole nationalism* that aimed for recognition and participation in the political mainstream. Imbued by contemporary liberal ideology and advocating for a gradual separation from Spain, he used the power of the plumed pen by writing novels to depict Spanish tyranny and immorality (Rizal 1886, 1891). Although Rizal wrote in Spanish and German, his translated works became metaphors for Filipino oppression and transformed revolutionary yearnings into actions. Stirred by Rizal's writings to put an end to the Spanish regime in the Philippines, the largely peasant, secret fraternal organization *Katipunan* (translated as revolutionary brotherhood) led by Andres Bonifacio fought with bolos and guns. Bonifacio's *vernacular nationalism* drew from the signs and beliefs of local and indigenous cultures to heighten consciousness and strengthen solidarity (Ileto 1979). With battles still raging and the country's future undecided, the revolutionaries declared independence on June 12, 1898, and established the republic of the Philippines.

The republic, unfortunately, did not last long (Storey and Lichauco 1926). While Spain fought to retain its control over the Philippine islands in the Pacific, it was also involved in a war against the United States in the Atlantic shores (Hoganson 1998). The Spanish-American War ended with the Treaty of Paris on December 10, 1898, in which Spain ceded control of the Philippines to the United States. Confronted by both pro-annexationists and anti-imperialists at home and in the archipelago, the United States had to respond to domestic and international demands that called for either the continuation of U.S. rule over the islands or the autonomy of the Philippines (Lanzar 1928; Stanley 1974). The U.S. government chose to retain the islands and instill a sense of *official nationalism* in order to create a transnational and multiethnic society while pursuing an imperialist agenda. U.S. President William McKinley's "Benevolent Assimilation" policy (Miller 1982), allegedly designed to democratize the Philippines and bring it under American tutelage primarily through education, was a way to stretch "the short, tight, skin of the nation over the gigantic body of the empire" (Anderson 1991, p. 86). In 1899, the United States officially annexed the Philippines and became the ensuing colonial master. However, the struggle for independence was not over. The Philippine–American War erupted (Tan 2002), fueled by *last wave nationalism* that drew from the three other models of nationalism (creole, vernacular, and official).

While Anderson's typologies of nationalism are suggestive for understanding national self-determination in the Philippines, they are also limited both in their explanatory power and by their Eurocentric bias. Even though they have influenced numerous studies of anticolonial movements, they have been criticized for privileging Western constructs that only perpetuate imperialist dynamics and mechanisms even after national independence. Subaltern Studies historian Partha Chatterjee, for example, takes issue with Anderson's formulation of last wave nationalism. He questions what the "postcolonial world" has left to imagine if Europe and the Americas, as "the only true subjects of history, have thought out on our behalf not only the script of colonial enlightenment and exploitation, but also that of our anticolonial resistance and postcolonial misery" (Chatterjee 1993, p. 5). He concedes that colonialism has produced destructive and beneficial consequences in the colonies. He asserts, nevertheless, that a certain part remains within the "natives" that is untouched by colonialism, a crucial piece that sparks and sustains authentic nationalist fervor. According to Chatterjee, imperialism partitions colonized societies into two distinct domains: a material or outside realm where the Western hegemony of money, politics, science, and technology reigns; and a spiritual or inside realm that bears precolonial and indigenous cultural values. Anticolonial nationalism develops, he argues, in the second domain where revolutionary movements thrive, sanitized from Western contamination.

Whereas Anderson's typologies situate the colonized as identifying with Eurocentric models, Chatterjee's spiritual concept of nationalism locates them as counter-identifying with indigenous constructions. The process of identification (with the West) and counter-identification (with the natives) produces a binary opposition that maintains a particular version of nationalism as a conservative/accommodationist stance and the other as liberal/oppositional. The colonized are thus limited to the choice between only two options. This opposition also raises questions about "purity" in colonial situations (Bhabha 1994; Stoler 1995). The colonizers' anxiety about their purity is related to their capacity to retain a dominant and thereby dominating position. The apprehension of the colonized, on the other hand, is connected to their ability to hold on to their indigenous cultures, stop further colonial impositions, and have options for self-determination. To be demonstrated in the next sections through the experiences and perspectives of Camilo Osias, an educator who straddled between U.S. imperialism and Filipino sovereignty, neither the dichotomy of identification and counter-identification nor the insistence for purity is helpful. Situated at the crossroads of changing colonial regimes and nationalist uprisings, Osias embodied and performed "disidentifying nationalism," an especially insightful framework to understand the issues, problems, and patterns in identity formation, anti-imperialist resistance, and revolutionary pedagogy.

REVOLUTION AS PEDAGOGY

Camilo Osias was born on March 23, 1889 from peasant parents, in the "typical" town of Balaoan, La Union, in northern Philippines where its approximately 8,000 inhabitants lived simply and parsimoniously (Osias 1971). The sixth out of eight children (four of whom died in infancy), he planted in the rice fields, grew vegetables, tended livestock, and fished to help his family obtain food and money. At an early age, Osias already discovered and understood the harsh realities of loss, poverty, and deprivation. His first teachers were his parents who taught him the alphabet, writing, and religious conviction. His parents' desire for their children to have a better future led them to enroll Osias in a private school where he learned Spanish grammar, Latin, geography, and mathematics. Osias and his classmates considered themselves better than the students in the town's newly established *escuela publica* that merely taught reading, writing, and the catechism. Since the public school system in the Philippines under the Spanish regime only began in 1863, the private schools, mostly under the supervision of Spanish friars, provided historically formal education to Filipino children and offered a more structured system of instruction (Alzona 1932; Bazaco 1939; Bureau of Education 1903, pp. 225–231; Osias 1917a).

Osias's tremendous capacity to memorize and do well under pressure garnered him quite a favorable standing in his hometown. Proficient in Spanish and Latin, he won a grammar competition held in the central plaza. Consequently, local officials selected him to deliver the welcome speech to a high-ranking Spanish judge at an inaugural celebration in the province capital. His impressive performance at the contest and at the ceremony earned him the reputation of being a diligent and bright student. No more than eight years of age, he taught reading, writing, math, catechism, and grammar to 12 to 14 kids around his age, a few even older than him, whose parents paid him as a tutor. His accomplishments brought him honor and family pride but did not make him arrogant; in fact, they made him more persistent and studious. However, the townspeople's talk about their "Balaoan boy [becoming] a La Union boy" foreshadowed Osias's ascendance to greater prominence and his exposure to more cosmopolitan ideas and settings (Osias 1971, p. 33).

In 1896, Osias's formal schooling was interrupted and his education under revolutionary times commenced. The uprising against the Spanish rule and for Philippine independence broke out on August 29 of that year when Andres Bonifacio and the *Katipuneros* declared war on Spain (Karnow 1989). The fighting between the Spanish military and the Filipino guerilla "insurgents" reached Balaoan in a few months, and the colonial officials shut down the schools and used them as headquarters. The Spaniards incarcerated or executed those suspected of being sympathetic to the guerillas and forced able-bodied men, such as Osias's older brother, to enlist and join the Spanish forces. Osias's father fled to the mountains upon learning that the Spanish parish priest wanted him captured for helping the townspeople air their grievances. From his mother, Osias found out about the garroting of three Filipino priests, Fathers Gomez, Burgos, and Zamora, who challenged Spanish theocracy. From his uncle who served as a *Katipunan* lieutenant, he learned about the principles of this fraternal organization and translated them from Tagalog to Ilokano[1] to inspire others to fight for their country and liberty. Suspicious of misleading information, he led his friends to drive away the Spanish *bandillo* or town crier who brought news of battles with only Filipino casualties. Osias's home and the streets, therefore, functioned as teaching sites that initiated his informal education on the nationalist struggles against colonialism. The revolution became a form of pedagogy that instilled in him a critical perspective on Western authority and an ardent passion for Filipino autonomy.

When the fighting subsided, Osias's deeply religious mother sent him to the larger town of Vigan, Ilocos Sur, to study in a seminary in order to fulfill her ambition to have a son educated for priesthood. Although his mother saw priesthood as a noble calling and a family blessing, he was aware of the material affluence and political influence of Spanish friars who made themselves "Little Gods" (Osias 1971, p. 57). When the revolution against Spanish control ended in 1898, it turned into another war, this time against the Americans (Hoganson 1998; Miller 1982; Storey and Lichauco 1926; Tan 2002). Filipinos felt betrayed that Americans, whom they thought came to help them oust the Spanish colonizers, were interested in keeping the Philippines. The U.S. army came to Vigan and closed all schools, thereby halting once again Osias's schooling. His informal lessons on revolution and nationalism, nonetheless, continued. As a complicit collaborator in the more overtly armed political mobilizations of the Filipino guerillas, young Osias served as an informant, message carrier, and lookout. He utilized the subaltern strategies

of resistance (Scott 1985) by befriending American soldiers to acquire information, employing gossips to seek support from local villagers, and participating in covert meetings. The ways in which he mobilized his formal and
informal education to subversively aid the propagation of revolutionary consciousness and the Filipino troops fighting the Spaniards and the Americans
clearly signaled the beginning of disidentifying nationalism that opened new
spaces of viable empowerment and resistance for the colonized.

EDUCATION IN THE COLONY
AND THE METROPOLIS

Although education was part of the American program to bring literacy and
progress to the country and to encourage people's acceptance of the new
colonial order, Filipinos utilized it for their own nationalist agenda. In 1901,
the 12-year-old Osias returned to his hometown and resumed his education
under U.S. tutelage. The American soldiers sparked his interest, and he
began to jot down phrases that he heard, such as " 'Hoyogon?' (Where are
you going?) 'Hoyocom?' (Where did you come from?) and 'Hislocanachi'
(Here's looking at you)" (Osias 1971, p. 68). A lieutenant noticed him taking
notes and organized an English class for him and his friends, which Osias
later recalled in these terms:

> Day after day, after the morning drill, the soldier teacher wet with perspiration
> came to the class, and taught us our first English lessons. We learned fast from
> our teacher and from watching the soldiers play ball. We picked up words like
> ball, bat, run, catch, out, etc. (Osias 1971, p. 69)

Many U.S. soldiers initiated the first American classrooms and schools in
the archipelago during and after the Philippine-American War (Bureau of
Education 1901; Gates 1973). Eventually civilian educators, most of whom
had normal training or undergraduate degrees, replaced the soldiers (Bureau
of Education 1902, pp. 206–216, 219–223; 1905, pp. 750–753; Freer 1906;
Lardizabal 1991; Racelis and Ick 2001; United States Embassy 2001). One of
these teachers, William Rosenkrans of New York came to Balaoan, La Union,
and established regular day classes and an evening class to train teachers how
to instruct beginners (Bureau of Education 1902, p. 214; 1905, pp. 869–872).
Osias enrolled in both and, after a few weeks, was asked by Rosenkrans to teach
a beginner's class. Within a short period of time, Osias went from learning
English by listening to soldiers in their conversations and games to teaching
English to other students. The change from elitist Spanish-style instruction
to mass-oriented American schools was a shift not only to a new system of
education, but also in the approach to neocolonial culture and participation.
Within the American regime, Osias would find new ways to navigate through
the educational system that promoted individual advancement and the general development of the Philippines.

Whereas education under the Spanish administration was reserved for the
exclusive few, the United States introduced its common school model in the

archipelago and created a more extensive system of schooling that was made available to more children. From 1902 to 1910, the number of schools climbed from an estimated 2,000 to 4,581 and the average daily attendance rose from roughly 150,000 to over 450,000 pupils (Bureau of Education 1910, p. 306). The American general superintendent of schools in the Philippines proudly declared at the first division superintendents' convention in 1903:

> [s]hould the work of education in these islands prove successful, there will be no brighter page in American history than that which tells the tale of the enlightenment and uplifting of a downtrodden people[,] than that which recounts the fact that an humble people was taken from the customs of three hundred years, placed upon the highways of progress, and prepared to take its place upon a plane of equality with the other peoples of the earth. (Bureau of Education 1903, p. 563)

Such benevolence purposely deployed dichotomous representations of two strangers on the Philippine shores: the Americans as modernizing and enlightening saviors and the Spaniards as old-fashioned and oppressive theocrats. Whereas Spanish colonialism was tied to the cross and the sword as allegories of the religious and military establishments, the Americans utilized the book as the hegemonic symbol of education to acquire people's acquiescence. In this crafted portrayal, the United States opened the gates of education, ushering in literacy, democracy, and "civilization" to Filipinos.

American education undoubtedly opened new opportunities to Camilo Osias. When Rosenkrans moved to La Union's capital of San Fernando to set up the provincial high school, he asked Osias to continue his studies there. Perhaps recognizing his passion for learning and teaching, Osias's parents allowed him to go. In the American-run high school, as recalled by Osias:

> [t]he classes were not very well organized, but we had different subjects like Grammar, Arithmetic, Geography, Civics and History, Music. I liked Grammar which was in some ways like my *gramatica* in Spanish. I shone in parsing, in the different parts of speech, in conjugation, and in diagramming. (Osias 1971, p. 70)

The newly established high school of La Union suffered from initial disorganization and minimal resources (Bureau of Education 1903, pp. 365, 468; 1904, pp. 618, 633; 1905, p. 871). Compared with its Spanish counterparts, however, the American school offered new and more courses. While Osias excelled in grammar and language due to his strong foundation in Spanish, he also began to identify with his American teachers and their dispositions.

> My teachers were all good. I liked to see them come to school everyday in clothes clean and well pressed and shoes well-polished. I admired them and my ambition was to be a high school teacher. (Osias 1971, p. 70)

After meeting the principal and the division superintendent and realizing that they were in charge of teachers and schools, he decided to forego his initial ambition of becoming a teacher and, instead, chose to become an administrator. According to an American professor analyzing the Philippines at the time, education played a crucial role in Filipino participation in the U.S. colonial regime since it was the key to their preparation to pass the civil service examination, which opened doors to government service (Willis 1905). Performing well in the American high school and setting his career goal to educational service were Osias's strategic moves not only to be competitive in the job market, but also to be in a position of authority that could push for changes in the system. In order to accomplish his goals, Osias realized that he had to learn from and master the American curriculum and pedagogy.

In 1905, Osias took and passed a rigorous test that assessed his knowledge of English grammar, geography, U.S. history, arithmetic, and physiology, and became one of the select few who was sent to study in the United States (Bureau of Education 1904, pp. 680–683; 1905, pp. 791–794). Through the *pensionado* program established by the governing Philippine Commission, cohorts of first-rate Filipino students became government-sponsored scholars that pursued higher education abroad (Carpio 1934; Olivar 1950; Osias 1925). While the program's initial goal was to train Filipino teachers infused with a comprehensive knowledge of American life, education and government that could be transplanted to their native country, it expanded the scope to include the fields of agriculture, engineering, business, and medicine (Bureau of Education 1901, p. 7; 1905, p. 797). Demographically, the *pensionados* were mostly men from well-to-do families with political connections to the colonial government as well as financial resources to enroll their children in superior schools. Although the first cohort of 1903 had the largest contingency of 102 *pensionados*, by 1910 the Philippines had sent a total of 207 students to the United States, eight of whom were women (Bureau of Education 1910, p. 297).

When Osias expressed interest in the area of pedagogy, the superintendent of Filipino students in America, William Sutherland, enrolled him in Western Illinois State Normal School along with five other Filipinos. Osias received his normal diploma at Western Illinois in 1908 and took summer courses at the University of Chicago. He then completed a bachelor's degree in Education and a certificate in School Administration and Supervision at Teachers College of Columbia University in 1910. Years later, both institutions accorded him their highest honors. The March 3, 1965 issue of *Western Courier* showcased Osias as the recipient of Western Illinois' first Distinguished Alumnus Award, and during Columbia University's One hundred and seventy fifth anniversary celebration in 1929, Osias received the University Medal for Public Service (Miller 1929). These awards testified to the considerable impact that Osias had on American people and institutions, affirming that imperial encounters and influence between Filipinos and Americans were, after all, a two-way street.

Osias's five years in the United States were extremely productive and enriching. He spent many hours in the libraries and often slept only three to four hours each night. One of his professors remarked that he "worked so hard as to memorize the lessons on Shakespeare and the commentaries" (quoted in Osias 1971, p. 80). At Western Illinois, Osias was highly involved in extra-curricular activities. He regularly attended Protestant church services and was chairman of the campus YMCA, a significant change of religious affinity rooted in his criticism of Spanish Catholicism and his interest in American Protestantism that started in the Philippines (Osias 1965; Osias and Lorenzana 1931). Various issues of the normal school's weekly newspaper *Western Courier* highlighted Osias as an active member of the Platonian literary society who competed in debates and oratorical contests. In the 1908 yearbook *Sequel*, he was dubbed "The Patrick Henry of the Philippines" who "talks like thunder and sings like a nightingale." In addition, his versatility ranged from performing in a German comedy and singing with a Filipino musical quartet to being the tennis doubles champion and the substitute quarterback in football. Like many *pensionados* who saw themselves as their country's representatives and whose education, demeanors, and small number did not pose a threat to the majority, he and the other Filipino students were noted for their "good manners and faithful attention to their work" and were welcomed by their American peers (Black 1905, p. 24). In Illinois, he heard Booker T. Washington give a talk, which inspired and guided his eventual policies and programs in industrial and rural education in the Philippines (Osias 1921). At Teachers College in Columbia University, he took courses from leading progressive scholars, such as John Dewey, Paul Monroe, Edward Thorndike, David Snedden, and George Strayer (his adviser), whom he credited as "excellent teachers and professors" that "left an indelible imprint of their ideas and principles upon my life and character" and "helped cultivate my passion for my faith in and my devotion to education" (Osias 1971, p. 110; 1914; 1940; 1954).

Prior to his graduation in New York and in preparation for civil service in the Philippines, Osias took the superintendency exam that had been only reserved for Americans and became the first Filipino to pass it. His decision to take the administrative test punctuated his shift in attitude toward American culture in general and education in particular. Even though he desired to learn from and utilized his American education in the Philippines and in the United States, he began to disengage from this dominant structure during his stay in the United States. He demonstrated his increasing dissatisfaction with the colonial order that continued to subjugate him and the people of his country through his delivery of the highly patriotic speech "The Aspiration of the Filipinos," which called for his country's independence, and through his rejection of the teacher civil service exam that was relegated by Americans to Filipinos. Both actions emphasized his awareness of the unequal power dynamics in Filipino–American imperial encounters and his ability to assert choices that resisted further cooptation. His identification with U.S. ideals and lifestyle juxtaposed with his counter-identification with the oppressed conditions in the Philippines produced a disidentifying nationalism that worked within and against the American system to bring about progressive change for his country.

As a hybrid approach to deconstructing the majority's language, power, and operations, disidentifying nationalism serves as a bridge between Western and indigenous affinities. It invokes a third space that avoids the pitfalls of the either/or strategy and deploys the *potential* usefulness of the both/and as a revolutionary tool to bring together seemingly contradictory elements in moments of tension and ambivalence. According to performance studies scholar José Esteban Muñoz:

> disidentification scrambles and reconstructs the encoded message of a cultural text in a fashion that both exposes the encoded message's universalizing and exclusionary machinations and recircuits its workings to account for, include, and empower minority identities. (Muñoz 1999, p. 31)

Operating as a three-part mechanism undergirded by the notion that culture can be read and interpreted like a text, disidentification first names and specifies the symbols and codes of cultural materials. These materials, such as policy documents, school textbooks, photographs, or people's outfits and behaviors, are resources and representations that offer insights regarding an individual, institution, or community within particular sociohistorical contexts. It then unpacks the meanings of these cultural symbols to reveal the ways in which they privilege majority values and marginalize minority perspectives. Finally, it reconfigures and recycles these codes in order to represent "a disempowered politics or positionality that has been rendered unthinkable by the dominant culture" (Muñoz 1999, p. 31).

In potentially revolutionary moments, disidentifying nationalism provides a way for minority subjects to assert their agency. What has been unthinkable to the people in power becomes a source of hope and justice to the oppressed. Within anticolonial movements, disidentifying nationalism valorizes the values and influences of neither the metropole nor the periphery. It appropriates from both, in different degrees, at various times, and for multiple strategic purposes. Osias displayed his commitment to the struggle for Filipino self-determination within the imperial encounters of pedagogical engagements in his colonized country and in the United States. These encounters in classrooms, extra-curricular activities, and even civil service exam, had tremendous revolutionary significance since they provided the arenas for Osias to enact his nationalist resistance and set the foundation for his career in education and politics.

From Potentiality to Actuality

Navigating the complex and often problematic in-between spaces within imperial encounters is akin to moving between a rock and a hard place. The return of the *pensionados* to the Philippines was greeted by a mixture of awe, admiration, intrigue, and envy. Even though they were recruited and placed in white-collar employment and accorded a special status, they also faced sociocultural and professional challenges. Osias became "somewhat of a curiosity" in his hometown (Osias 1971, p. 123). People referred to him as an "American boy" partly due to his wearing a long woolen suit in such a tropical and agricultural place. According to a daughter of another *pensionado*, the

American-educated scholars also faced "race prejudice" from white Americans (Olivar 1950, p. 83). For instance, within a few weeks of teaching in San Fernando, Osias was disliked by some of his American colleagues since the American principal set him as an example to be followed by the entire staff. Regardless how hard he worked, however, he was not compensated at the same rate as American teachers. With a bachelor's degree and an administration certificate from Columbia University and a civil service qualification for superintendency, he received a salary of 1,080 pesos a year. By contrast, an American teacher with only a normal education or a bachelor's degree was paid 4,000 pesos a year. Osias, like many *pensionados*, occupied a precarious position and had to prove to both Filipinos and Americans that he was still a Filipino and as good as any American in order to earn their trust and confidence.

Despite his initial setbacks, Osias immediately rose in the educational ranks and gained prominence among American and Filipino officials. Within the first three years, he became a supervisor of teachers in San Fernando, San Juan, and Bacnotan, and then an academic supervisor in the country's capital of Manila with the largest city school system. In 1915, at the age of 26, he accomplished another breakthrough by becoming the first Filipino division superintendent of schools in the Philippines. After two years as a superintendent of Bataan, Mindoro, and Tayabas, he was promoted to an assistant directorship in the Bureau of Education, thereby becoming the highest-ranking Filipino in the department. His meteoric rise testified to his superb competence as a teacher and administrator and his exceptional ability to work through the tensions of the U.S. colonial project. These tensions included the tremendous growth in the number of public schools and attending students, the high demand for trained education personnel, and the persistent agitation for the country's independence and the Filipinization of the public sector. American trained yet still a Filipino nationalist, Osias worked within and against the dominant structures of colonial education, which bolstered his personal and professional standing as well as the advancement of Filipino careers, politics, and sovereignty.

The philosophy of "dynamic Filipinism" formed the foundation of Osias's educational and political praxis. This "intelligent and constructive patriotism" sought "to preserve and develop what is best in Philippine culture, civilization, and philosophy, and to graft on them the best that is foreign if this grafting can be accomplished advantageously" (Osias 1940, pp. 52–53). In his 1921 inaugural speech as the first president of the National University, he elaborated on the compatibility of nationalism and internationalism. According to Osias, if the aim of education was "to secure for humanity as a whole and for every human being the highest and fullest measure of freedom, happiness, and efficiency," then Filipino education must serve as "an agency of harmonizing the cultures and civilizations of the East and of the West." He called for "a Filipinism that is compatible with world progress" as "a foundation upon which the superstructure of a new humanity shall rest." He abhorred the "traditional policy of 'splendid isolation' " since people and cultures, he contended, were "interrelated and interpenetrating." He encouraged a "sane Filipinization" that was by "no means an anti-foreign movement," but one embedded in "civic responsibility" and "world consciousness" (Osias 1926, pp. 1–20).

Osias's dynamic Filipinism manifested in the transformation of educational curriculum, pedagogy, and administration. As a curriculum pioneer, he wrote the *Philippine Readers* series for grades one to seven that contained Filipino and Western stories, folklores, biographies, and historical events (Osias 1927; 1932a; 1932b; 1932c; 1932d; 1932e; 1959). Although the first edition was published in the early 1920s, subsequent editions became affectionately known as "Osias Readers" to honor the first Filipino author of Philippine school textbooks. As an innovative pedagogue, he showcased model classrooms and recitations, conducted training institutes for veteran teachers, created a manual on *Methods and Practical Suggestions* for novice ones, and pioneered a new course on "Good Manners and Right Conduct" (Bureau of Education 1913; Osias 1914). As a community leader, he wrote *Barrio Life and Barrio Education* to address persistent literacy and vocational problems in the rural areas (Osias 1921).

In addition, he fostered the spirit of alliances without compromising his desired goals. He wrote a manual for administrators that instructed them how to supervise teachers in order to provide culturally relevant curriculum and how to work with municipal officials in order to acquire school sites and construct buildings (Osias 1918). He also established schools for the indigenous groups of Negritos in Bataan and Mangyans in Mindoro. When he was in charge of the "non-Christian" provinces, he pursued a "policy of amalgamation" that intended to unite and cultivate understanding among Christians, Muslims, and other local spiritual communities (Osias 1971, p. 151). Grounded in the notion of unified pluralism, Osias's disidentifying nationalism emerged in his educational praxis that connected various communities together and celebrated the rich diversity of the Philippine nation.

When Osias became the first Filipino division superintendent of schools, he worked to Filipinize the entire educational staff. In charge of the first all-Filipino division, he knew that he had to make this experiment a "real success" since Filipino hopes and American ambivalence centered on this "Bataan Republic" (Osias 1971, p. 131). To demonstrate that a mixed staff could also work harmoniously under a Filipino superintendent, he opted for a combination of Filipino and American teachers and administrators in his Tayabas assignment (Osias 1917b). However, not everyone agreed with the Filipinization policy. During an all-superintendents meeting, an American colonial official remarked, "Our [American] ways are superior and these people have to take them whether they like it or not" (quoted in Osias 1971, p. 138). Enraged by such arrogance and aware of his position as the sole Filipino representative in the room, Osias responded with "Americans are here to serve our people, and whosoever cannot sympathize with the customs and mores of the Filipinos has no right to be an educational official in these beautiful isles of the Pacific" (Osias 1971, p. 138). His educational leadership initiated and bolstered the Filipinization of educational administration and personnel since his promotion to higher positions led to the selection of mostly Filipino successors.

Osias's promotions opened various opportunities for his career development and nationalist desires. During a trip to the United States where he observed the latest educational practices, he joined the first Philippine Independence Mission to the U.S. Congress. For Osias's eloquent testimony to the joint Senate and House committee that oversaw American territories, Senator Warren Harding commented, "If you have half a dozen men like your Osias, you are entitled to your independence" (quoted in Osias 1971, p. 149). At that moment, the mere "*maestrillo*, an insignificant little teacher" became the "*Pambato de la Mision*, the pitcher or ace spokesman" (Osias 1971, p. 149). As prophesized by a Western Illinois classmate in the 1908 *Sequel* yearbook, his speech in the U.S. Congress would become his first of many. Elected senator to the Philippine legislature in 1925, Osias was then chosen by his colleagues to be a Resident Commissioner to the United States in 1929. With a non-voting seat in the U.S. House of Representatives, he fought for six years for his country's freedom. From April 1932 to January 1933, he lobbied American legislators and gave compelling speeches to, first, pass the Hare-Hawes-Cutting Bill, known as the Philippine Independence Bill, and then successfully override President Herbert Hoover's executive veto (Osias and Baradi 1933). This bill preceded the Tydings-McDuffie Act of 1934 that eventually established a commonwealth period as a transition from colonial rule to complete political sovereignty. After World War II, the aspiration of Camilo Osias and other Filipinos became a reality; their beloved country became independent on July 4, 1946. Although Osias continued an illustrious career in politics that spanned until the late 1960s to serve the Filipino people in the domestic and international spheres as a senator, ambassador, and diplomat, one major political gem eluded him; the presidency of the Republic of the Philippines (Bananal 1974). However, as a leader in the Filipinization movement, as the father of the Philippine modern educational system, and as one of the country's greatest statesmen, Osias left an indelible mark that could be surpassed by only a few in the history of the Philippines.

LESSONS FROM A FILIPINO NATIONALIST

The legacy of Camilo Osias lives not only through the sovereignty of the Philippines and the transformation of Filipino education in the early twentieth century, but also through his embodied pedagogy of revolution. Raised in the tumultuous era of wars and uprisings and shaped by imperialist education in his home country and the United States, Osias constructed and performed an oppositional hybrid identity that derived from his multiple experiences and backgrounds. His disidentifying nationalism was neither complete conformity to the American colonial agenda nor an invocation of separatist and essentialist nativism. It reconfigured instead the contacts with and influences of peasants, clergymen, teachers, and political leaders. His formal and informal training came from various pedagogical settings, such as Philippine and American classrooms, guerrilla actions, oratorical contests, civil service exams, publication of books and other manuscripts, and the halls of the U.S. Congress.

Through his schooling and career in education and politics, Camilo Osias constructed and performed a disidentifying nationalism that drew from both Filipino and American cultures, that did not support narrow and chauvinistic attitudes, and that refashioned dominant ideas and structures in order to produce what Edward Said has called "a new humanity" that uplifts all, especially the oppressed and colonized. Working in conjunction with political, sociocultural, and subaltern strategies of resistance, the significance of disidentifying nationalism is the development and proliferation of in-between spaces and hybrid identities that infiltrate, modify, and subvert mainstream discourses and structures by employing their representational and material systems against themselves. Strategically troubling and bringing into play the tensions of the colonial project, these spaces and identities focus on the dominant codes, dismantle their conventional meanings, and extend their use in order to empower the colonized and advocate for marginalized politics. They insist on "the right to see the community's history whole" in order to "[r]estore the imprisoned nation to itself" and to reconceptualize resistance not solely as a reaction to imperialism but as a more integrative view of human history (Said 1993, pp. 215–216).

By appropriating from and working through the tensions of dominant and subordinated cultures, the colonized can *move beyond* the dichotomy of identifying with the West and counter-identifying with the native. The oppositional power of disidentifying nationalism generates a critical awareness and praxis that marshal yet undermine the dominant culture. It also seeks to create a viable and empowering alternative that honors indigenous traditions and attends to local needs and concerns. In the context of imperial encounters and other situations with unequal power relations, revolutionary individuals and groups modify lessons learned "from the top" and adapt to the changing conditions in order to advocate for and with those "at the bottom." Honing on the fluidity and contingency of power dynamics and identity formations, disidentifying nationalism ultimately demonstrates the revisioning of transnational history and colonial agency as well as the emergence of new revolutionary and pedagogical practices.

Notes

My deep appreciation goes to Judy Wu and Kenneth Goings for reading early versions of this paper, to Luz Calvo for our discussions on the theory of disidentification, to Margaret Mills and Shelly Wong for organizing the Revolution and Pedagogy conference where I presented a version of this paper, and most especially to Tom Ewing for his keen insights and patience.

1. In an archipelago with over 7,100 islands and over 80 different dialects, the question of national language was in the early 1900s, and remains to this day, a highly contested issue. The majority of the people living in the country's capital of Manila and the surrounding central-southern plains of Luzon, where the Katipuneros

initially organized, speak Tagalog. The northern part of Luzon, the Ilokos region and La Union, where Osias was born and his family lived, predominantly uses Ilokano.

REFERENCES

Agoncillo, Teodoro A. 1990. *History of the Filipino People*. 8th ed. Quezon City, Phils.: Garotech.

Alidio, Kimberly A. 2001. *Between Civilizing Mission and Ethnic Assimilation: Racial Discourse, U.S. Colonial Education, and Filipino Ethnicity, 1901–1946*. Ph.D. diss., University of Michigan.

Alzona, Encarnacion. 1932. *A History of Education in the Philippines, 1565–1930*. Manila: University of the Philippines Press.

Anderson, Benedict. 1991. *Imagined Communities: Reflections on the Origin and Spread of Nationalism*. Revised ed. London and New York: Verso.

Bananal, Eduardo. 1974. *Camilo Osias: Educator and Statesman*. Quezon City: Manlapaz.

Bazaco, Evergisto. 1939. *History of Education in the Philippines, Vol. 1: Spanish Period—1565–1898*. Manila: University of Santo Tomas Press.

Bhabha, Homi K. 1994. *The Location of Culture*. London and New York: Routledge.

Black, Plummer. 1905. "The Filipinos." *Western Courier* November 15.

Bureau of Education. 1901. *First Annual Report: Department of Public Instruction*. Reprint, Manila: Bureau of Printing, 1954.

———. 1902. *Second Annual Report: Department of Public Instruction*. Reprint, Manila: Bureau of Printing, 1954.

———. 1903. *Third Annual Report: Department of Public Instruction*. Reprint, Manila: Bureau of Printing, 1954.

———. 1904. *Fourth Annual Report: Department of Public Instruction*. Reprint, Manila: Bureau of Printing, 1954.

———. 1905. *Fifth Annual Report: Department of Public Instruction*. Reprint, Manila: Bureau of Printing, 1954.

———. 1910. *Tenth Annual Report: Department of Public Instruction*. Reprint, Manila: Bureau of Printing, 1957.

———. 1913. *Good Manners and Right Conduct: For Use in Primary Grades*. Manila: Bureau of Printing.

Cabral, Amilcar. 1994. "National Liberation and Culture." In *Colonial Discourse and Post-Colonial Theory: A Reader*. Patrick Williams and Laura Chrisman, eds. New York: Columbia University Press, pp. 53–65.

Campomanes, Oscar V. 1995. "The New Empire's Forgetful and Forgotten Citizens: Unrepresentability and Unassimilability in Filipino-American Postcolonialities." *Critical Mass: A Journal of Asian American Cultural Criticism* vol. 2, no. 2, pp. 145–200.

Carpio, Remigia D. 1934. *A Study of the Philippine Pensionado System Abroad (1903–1928)*. Master's thesis, University of the Philippines.

Chatterjee, Partha. 1993. *The Nation and Its Fragments: Colonial and Postcolonial Histories*. Princeton: Princeton University Press.

Choy, Catherine Ceniza. 2003. *Empire of Care: Nursing and Migration in Filipino American History*. Durham: Duke University Press.

Constantino, Renato. 1966. *The Filipinos in the Philippines and Other Essays*. Quezon City: Malaya.

Cooper, Frederick and Ann Laura Stoler, eds. 1997. *Tensions of Empire: Colonial Cultures in a Bourgeois World*. Berkeley and London: University of California Press.

Espiritu, Augustu Fauni. 2000. *"Expatriate Affirmations": The Performance of Nationalism and Patronage in Filipino-American Intellectual Life*. Ph.D. diss., University of California, Los Angeles.

Fanon, Frantz. 1963. *The Wretched of the Earth*. Translated by Constance Farrington. New York: Grove.

Freer, William Bowen. 1906. *The Philippine Experiences of an American Teacher*. New York: Charles Scribner's Sons.

Fujita-Rony, Dorothy. 2003. *American Workers, Colonial Power: Philippine Seattle and the Transpacific West, 1919–1941*. Berkeley: University of California Press.

Gates, John Morgan. 1973. *Schoolbooks and Krags: The United States Army in the Philippines, 1898–1902*. Westport: Greenwood.

Guha, Ranajit. 1982. "On Some Aspects of the Historiography of Colonial India." *Subaltern Studies I: Writings on South Asian History and Society*, pp. 1–8.

Hoganson, Kristin L. 1998. *Fighting for American Manhood: How Gender Politics Provoked the Spanish-American and Philippine-American Wars*. New Haven: Yale University Press.

Ileto, Reynaldo C. 1979. *Pasyon and Revolution: Popular Movements in the Philippines, 1840–1910*. Quezon City: Ateneo de Manila University Press.

Jacobson, Matthew Frye. 2000. *Barbarian Virtues: The United States Encounters Foreign Peoples at Home and Abroad, 1876–1917*. New York: Hill and Wang.

Kaplan, Amy and Donald E. Pease, eds. 1993. *Cultures of United States Imperialism*. Durham: Duke University Press.

Karnow, Stanley. 1989. *In Our Image: America's Empire in the Philippines*. New York: Random House.

King, C. Richard, ed. 2000. *Postcolonial America*. Urbana: University of Illinois Press.

Lanzar, Maria C. 1928. *The Anti-Imperialist League*. Ph.D. diss., University of Michigan.

Lardizabal, Amparo Santamaria. 1991. *Pioneer American Teachers and Philippine Education*. Quezon City: Phoenix.

May, Glenn Anthony. 1980. *Social Engineering in the Philippines: The Aims, Execution, and Impact of American Colonial Policy, 1900–1913*. Westport: Greenwood.

McClintock, Anne. 1995. *Imperial Leather: Race, Gender and Sexuality in the Colonial Contest*. New York: Routledge.

Miller, Clyde R. 1929. "Columbia University Observes its One Hundred and Seventy-Fifth Anniversary." *Teachers College Record* (November), pp. 245–261.

Miller, Stuart Creighton. 1982. *"Benevolent Assimilation": The American Conquest of the Philippines, 1899–1903*. New Haven: Yale University Press.

Muñoz, José Esteban. 1999. *Disidentifications: Queers of Color and the Performance of Politics*. Minneapolis: University of Minnesota Press.

Olivar, Celia Bocobo. 1950. *The First Pensionados: An Appraisal of their Contribution to the National Welfare*. Master's thesis, University of the Philippines.

Osias, Camilo. 1908. "The Aspiration of the Filipinos." *Western Courier* May 28.

———. 1914. *Syllabus on Educational Methods and Practical Suggestions for Teachers*. Manila: Fajardo's Printing.

———. 1917a. *Education in the Philippines under the Spanish Regime*. Manila: Philippine Education.

———. comp. 1917b. *Division of Tayabas Circulars*. n.p.

———. 1918. *Notes on Supervision*. Manila: n.p.

Osias, Camilo. 1921. *Barrio Life and Barrio Education*. New York: World Book.

———. comp. 1925. *Philippine School Laws, 1900–1925*. Manila: n.p.

———. 1926. *Our Education and Dynamic Filipinism: Speeches and Materials*. Manila: n.p.

———. 1927. *The Philippine Readers, Book 1*. Boston: Ginn and Co.

———. 1932a. *The Philippine Readers, Book 2*. Revised ed. Boston: Ginn and Co.

———. 1932b. *The Philippine Readers, Book 3*. Revised ed. Boston: Ginn and Co.

———. 1932c. *The Philippine Readers, Book 4*. Revised ed. Boston: Ginn and Co.

———. 1932d. *The Philippine Readers, Book 5*. Revised ed. Boston: Ginn and Co.

———. 1932e. *The Philippine Readers, Book 7*. Revised ed. Boston: Ginn and Co.

———. 1940. *The Filipino Way of Life: The Pluralized Philosophy*. Boston: Ginn and Co.

———. 1954. *Life-Centered Education*. Quezon City: Bustamante.

———. 1959. *The Philippine Readers, Book 6*. Revised ed. Boston: Ginn and Co.

———. 1965. *Crusade for the Separation of Church and State*. Manila: National Senate.

———. 1971. *The Story of a Long Career of Varied Tasks*. Quezon City: Manlapaz.

Osias, Camilo and Mauro Baradi. 1933. *The Philippine Charter of Liberty*. Baltimore: French-Bray.

Osias, Camilo and Avelina Lorenzana. 1931. *Evangelical Christianity in the Philippines*. Dayton: United Brethren Publishing.

Racelis, Mary and Judy Celine Ick. 2001. *Bearers of Benevolence: The Thomasites and Public Education in the Philippines*. Pasig City: Anvil.

Rafael, Vicente L. 2000. *White Love and Other Events in Filipino History*. Durham: Duke University Press.

Rizal, Jose. 1886. *Noli Me Tangere*. Reprint, translated by Camilo Osias, Manila: Asian Foundation for Cultural Advancement, 1956.

———. 1891. *El Filibusterismo*. Reprint; translated by Camilo Osias, Manila: Asian Foundation for Cultural Advancement, 1957.

Said, Edward W. 1993. *Culture and Imperialism*. New York: Vintage.

Salman, Michael. 2001. *The Embarrassment of Slavery: Controversies over Bondage and Nationalism in the American Colonial Philippines*. Berkeley: University of California Press.

San Juan, E. (Epifanio). 2000. *After Postcolonialism: Remapping Philippines-United States Confrontations*. Lanham: Rowman and Littlefield.

Schirmer, Daniel B. and Stephen Rosskamm Shalom, eds. 1987. *The Philippines Reader: A History of Colonialism, Neocolonialism, Dictatorship and Resistance*. Boston: South End.

Scott, James C. 1985. *Weapons of the Weak: Everyday Forms of Peasant Resistance*. New Haven: Yale University Press.

Singh, Amritjit and Peter Schmidt, eds. 2000. *Postcolonial Theory and the United States: Race, Ethnicity, and Literature*. Jackson: University Press of Mississippi.

Stanley, Peter W. 1974. *A Nation in the Making: The Philippines and the United States, 1899–1921*. Cambridge: Harvard University Press.

Stoler, Ann Laura. 1995. *Race and the Education of Desire: Foucault's History of Sexuality and the Colonial Order of Things*. Durham: Duke University Press.

———. 2001. "Tense and Tender Ties: The Politics of Comparison in North American History and (Post) Colonial Studies." *Journal of American History* vol. 88, no. 3, pp. 829–865.

Storey, Moorfield and Marcial P. Lichauco. 1926. *The Conquest of the Philippines by the United States, 1898–1925*. 1985 Reprint, Mandaluyong: Cacho Hermanos.

Tan, Samuel K. 2002. *The Filipino-American War, 1898–1913.* Quezon City: University of the Philippines Press.

United States Embassy. 2001. *To Islands Far Away: The Story of the Thomasites and their Journey to the Philippines.* Manila: Public Affairs Section.

Willis, Henry Parker. 1905. *Our Philippine Problem. A Study of American Colonial Policy.* New York: H. Holt and Company.

Wolfe, Patrick. 1997. "History and Imperialism: A Century of Theory, from Marx to Postcolonialism." *American Historical Review* vol. 102, no. 1, pp. 388–420.

Young, Robert J. C. 2001. *Postcolonialism: An Historical Introduction.* Oxford: Blackwell.

2

GENDER EQUITY AS A REVOLUTIONARY STRATEGY

COEDUCATION IN RUSSIAN AND SOVIET SCHOOLS

E. Thomas Ewing

In 1913, Moscow schoolteacher E. Kirpichnikova declared that her coeducational school "does not have the goal of eliminating the difference between the sexes, but instead seeks only to assist the natural development of the positive features of young people of both sexes." Drawing upon her nearly 20 year effort to increase girls' access to the schools, Kirpichnikova described how male and female pupils studied and played together. Yet, distinct patterns of behavior were still evident. During recess, boys ran around the courtyard while girls talked in small groups. In class, girls "conscientiously and punctually completed all assignments" yet boys "evaded subjects in which they had no interest." Recognizing these persistent differences, Kirpichnikova concluded that by allowing pupils to engage in "simple and comradely interactions," coeducation taught boys and girls to see in each other "not only a man or a woman, but also a human being *(ne tol'ko muzhchinu ili zhenshchinu, no i cheloveka)*" (Kirpichnikova 1914, pp. 75–88).

Kirpichnikova's argument thus embraced two seemingly contradictory beliefs. On the one hand, coeducational schools challenged gender roles by encouraging girls to participate in previously exclusive institutions. On the other hand, the experience of coeducation tended to confirm, perpetuate, and even reinforce "the difference between the sexes." Embedded in the stated goal of teaching pupils to see each new person as "not only a man or a woman, but also a human being" was the implied assertion that gender meant difference ("a man or a woman") as well as ("not only . . . but also") the overcoming of difference ("a human being"). For Kirpichnikova, how-ever, any sense of contradiction was resolved through "simple and comradely interactions" that educated girls and boys on the basis of common interests even as they prepared for their different roles as adults.

A quarter-century later, in 1939, Moscow educator A. Savich declared that the decision by the new Soviet revolutionary government in 1918 to make all schools coeducational had "torn out by the roots one of the most ridiculous prejudices from the past about the inequality *(neravenstvo)* of the mental facilities of a man and a woman." Reaffirming a common theme of Soviet discourse, Savich declared that only communism allowed for "real equality *(deistvitel'nogo ravenstva)* and mutual respect among men and women." But Savich also criticized educators who believed that coeducation was achieved simply by assigning male and female pupils to the same school. In fact, Savich asserted, coeducation was not intended to provide "an identical and equal education *(odinakovoe, ravnoe obuchenie)*" to boys and girls. If an "identical education" was the goal, this could just as easily be accomplished by mandating an identical curriculum for gender-separate schools. The real value of coeducation was "to develop real, friendly, and comradely relations between boys and girls based on the mutual recognition of full equality *(priznanie polnogo ravenstva)* first in school and then in life." The ultimate goal was to produce more acceptable, yet still gendered, patterns of behavior:

> Coeducation enables boys and girls to cooperatively influence each other, so that the more clearly expressed softness and emotionality of girls smoothes out the certain sharpness and at times even rudeness of boys, while the liveliness, vivacity, and enterprises of boys is transferred to the girls. The features of the two are converging, but this will not deprive either boys or girls of the characteristic and peculiar features of each sex. (Savich 1939, pp. 23–24)[1]

Writing at a time when coeducation was by law and in practice compulsory for 30 million Soviet pupils, Savich's article is strikingly ambivalent in its articulation of a form of equity that promises to overcome as well as preserve gender differences.

The tension between achieving equality and maintaining difference illustrated in these articles by Kirpichnikova and Savich suggests the need to examine schooling in terms of gender as a form of identity and a structure of relationships. In this context, gender refers to the meanings associated with perceived differences between the sexes and to the influence of these perceptions on behavior. Attention to gender thus requires moving "beyond" structural issues such as access and enrollment in order to examine "social relations" within and around the school (Heward 1999, p. 3). Rather than assuming that boys and girls in coeducational schools do and should receive the same education, this approach asks how the underlying assumptions made by advocates of coeducation actually perpetuate a gendered model of schooling. While focusing on two specific texts from the early-twentieth-century Russian and Soviet context, this chapter makes a broader argument about the limits and the potential of gender equity as a revolutionary strategy that connects directly to global patterns in this new millennium.

The terms "equity" and "difference" contain embedded tensions when used in reference to gender. In a highly influential argument, Joan Scott has used the tools of discursive analysis to challenge the dilemma posed to feminists in the "equality-versus-difference debate" with her assertion: "In fact, that antithesis itself hides the interdependence of the two terms, for equality is not the elimination of difference, and difference does not preclude equality." From Scott's perspective, therefore, the question is not whether Russian and Soviet schools were defined more by equality or more by difference, but rather to understand the tensions in the relationships between the two terms, to recognize "a more complicated historically variable diversity that is also differently expressed for different purposes in different contexts," and ultimately to recognize that difference itself is in fact part of the meaning of equality, which Scott defines as "the deliberate indifference to specific differences" (Scott 1988, pp. 38, 46–47).

Practical manifestations of these tensions in the United States can be seen in the changing debates on sex equity in schools. In the last decade, public discussion of sex bias has called into question the teleological assumption that more access to education automatically leads to gender equality (AAUW 1992; Orenstein 1994; Sadker and Sadker 1994; Tyack and Hansot 1992). Most recently, however, the terms of the debate have changed to recognize the contested meanings of the terms "equality" and "equity." In a 1998 report, the American Association for University Women drew this distinction between these two terms:

> Educational equity implies quality education and equal opportunities for all students . . . Equity differs from equality, which sets up a comparison, generally between two groups. If our concern were *equality*, the critical question would be whether students receive the *same* education. Equity poses a different question: do students receive the right education to achieve a shared standard of excellence? Although "equity" implies that students' educational performance and outcomes will be the same across groups of students, it does not imply that students need the same things to achieve those outcomes. (AAUW 1998, p. 3)

Echoing the distinction drawn by Scott, as cited above, this quote suggests that equity in schooling is not the same as identical schooling, and that differences in educational experiences are not the same as inequalities in educational access and achievement. Rather than focusing narrowly on measures to "fix" schools that "shortchange" girls (which had been the agenda of the path-breaking 1992 report by the same organization), this more recent document seeks to define and develop educational approaches that are most meaningful for all girls (and boys).[2] While the context was obviously different, the presence of this same underlying tension in the articles by Kirpichnikova and Savich provides a vantage point for asking how coeducational schools pursue this seemingly contradictory goal of balancing difference and equality.

These two articles are also significant because of their location in, but at opposite ends of, the Russian revolutionary context. Between Kirpichnikova's article in 1913 and Savich's article in 1939, two major, yet strikingly different, transformations had taken place. In 1917, the autocratic system ruled by Tsar Nicholas II was replaced by a Soviet system led by the

Bolshevik (later renamed the Communist) Party of Vladimir Lenin. In the years that followed, an ideology of autocracy, Orthodoxy, and nationality was displaced by promises of people's democracy, workers' control of production, and revolutionary internationalism, and a rigid system of social estates gave way to dramatic, and at times coerced, redistribution of property, resources, and privilege (Fitzpatrick 1994). Yet, the revolutionary impetus was difficult to sustain, and by the early 1920s, the Soviet government was forced to compromise its more egalitarian and democratic promises to retain power in the face of internal resistance and external hostility. The second revolution, this time a "revolution from above," began in the late 1920s under the leadership of Lenin's successor, Joseph Stalin. Warning of the danger of "capitalist encirclement," Soviet officials launched a campaign of economic development, including rapid industrialization and forced collectivization, that was designed to concentrate productive resources and political power in the hands of an increasingly authoritarian state. The ensuing social transformation, which culminated in mass terror in 1937–38, represented a continuation of processes begun in 1917, yet brought such far-reaching changes that it truly constituted a second revolution (Tucker 1990).While Soviet officials proclaimed the "realization" of the promises of the Bolshevik revolution, critics dubbed the increasingly hierarchical, authoritarian, and repressive Stalinist system a "betrayal" of revolutionary principles (Trotsky 1937, pp. 144–159). By 1939, when Savich's article was published, the Soviet Union was ruled by a dictator, terror was used against all suspected opponents, and social issues such as public education and women's equality served to glorify the "unprecedented achievements" of the world's only communist regime.

Elementary and secondary schools occupied a strategic location throughout this revolutionary transformation of modern Russia. In just over half a century, from the 1880s to the 1930s, enrollment increased from approximately one million to more than thirty million children, with the most dramatic increases coming in the 1930s (Ewing 2002, pp. 53–66). The Soviet regime openly embraced the revolutionary possibilities of the school. In 1918, Lenin declared: "The victory of the revolution can be consolidated by the school—the training of future generations will anchor everything won by the revolution" (cited in Lapidus 1978a, p. 135). The pedagogy of the Russian revolution thus followed, but with unique trajectories, the other political, social, and cultural transformations involved in the overthrow of the old system and the establishment of a new order. Historians have described in detail these attempts to "revolutionize" the school through administrative changes, curricular revisions, and instructional innovations (Holmes 1991). Revolutionary pedagogy was expected to have intensive effects by transforming the behavior, ideas, and very identity of pupils during and after their time in the schools. Whether the goal was the "moral preparation for life" promised by Kirpichnikova, the "victory of the revolution" anticipated by Lenin, or the "communist education" desire by Savich, Russian and Soviet educators shared a common belief that attitudes and behaviors learned in schools shaped the ways that individuals understood and responded to social structures and collective aspirations in Russia.

In terms of gender equity, however, the most "revolutionary" aspects of the Soviet educational system were the decisions in 1918 to make all schools coeducational, the efforts in the 1920s to reshape gender relations in and out of the school, and the rapid expansion in educational opportunities for girls that began in 1930.[3] Yet, gender equity serves as a critical measure of revolutionary transformations. In the decades before the revolution, the "emancipation" of women became a common demand of political movements from the liberals through the most radical socialists, although the relative significance attached to this objective varied considerably. The protests of working class women in February 1917 sparked the demonstrations that brought down the autocracy, yet women never again enjoyed as prominent a political role in the subsequent decades. Seeking to uphold Marxist-inspired promises to end the exploitation of the most oppressed social groups, the new Communist regime defined the "emancipation" of Russian women as one of its objectives, and Lenin promised that the new regime would "abolish all restrictions on the rights of women" (Lenin 1945, pp. 300–301). By the early 1920s, the new Soviet regime decreed equality in legal and civil rights for women, reformed marriage and divorce laws to expand women's options and protections, legalized abortion, and created a separate women's section of the Communist Party Central Committee (the *Zhenotdel*) to implement, enforce, and defend these actions (Clements 1994, p. 50).

Yet, this struggle for "emancipation" occurred not just on the level of state institutions and political ideology, but in the "everyday lives" of Soviet women at work, in the home, and in other social institutions. In many ways, the most fundamental changes in women's lives resulted not from the decrees and ideology of the regime, but from the social and economic transformations occurring in the First Five Year Plan (1928–32). In the decade that followed, the lives of Soviet women were transformed by processes of industrialization, urbanization, and political repression that expanded employment opportunities, increased educational access, constrained individual rights and personal liberties, and transformed relations within families, communities, and Soviet society as a whole (Bil'shai 1956, p. 130; Chirkov 1978, pp. 120–131; Lapidus 1978a, pp. 96–160; *Zhenshchina* 1936, pp. 3–35). At the same time, however, Soviet efforts to "emancipate" women exposed fundamental tensions in the revolution more generally. Was the goal to eliminate differences by making women the same as men, or was the goal to improve women's lives within distinct gender roles? The revolution that engulfed Russia thus exacerbated, but did not resolve, the underlying tensions shaping the discussion of coeducation.

The aspiration to gender equity and the persistence of gendered experiences in a revolutionary context are thus the focus of this chapter. The two texts by Kirpichnikova and Savich provide the main basis for evaluating the promise as well as the problems inherent in this revolutionary strategy. As a policy debate with the potential to affect all children enrolled in elementary and secondary schools, the issues raised in these articles were central to the lives of tens of millions of pupils over this quarter-century. More generally, this chapter asserts that studying the Russian/Soviet context demonstrates that achieving equity in education will not be complete without understanding the persistence of gendered relationship within the daily practices of schools. As the final section will argue, moreover, the tensions between gender and education revealed in the context of revolutionary Russia persist through to the present-day "postrevolutionary" global era.

COEDUCATION AND THE LIBERAL PROJECT

Kirpichnikova's argument for coeducation began with her personal background as a secondary-level teacher for the past 20 years. By making this statement on the first page, Kirpichnikova self-identified not as one of the very small number of educated women professionals in Russia, but also as someone defined by a lifetime of service to society. Kirpichnikova also openly identified with the other advocates and practitioners of coeducation, whom she described as a small band, "advancing gropingly into the darkness," desperate for any opportunity to "share personal experience with each other" (Kirpichnikova 1914, p. 78). In fact, Kirpichnikova's school was quite small: only three girls and twelve boys earned certificates at the first official graduation in 1913 (Kirpichnikova 1914, p. 77).

Yet, this commitment to expanding girls' access to schools was attracting considerable support across the political spectrum in early-twentieth-century Russia. By 1914, advocates claimed more than 150 coeducational schools in Russia (Sokolov and Tumin 1914, p. v). At the time of Kirpichnikova's article, both liberal and socialist wings of the opposition movement had endorsed the goal of promoting education for girls and women (Edmondson 1984, pp. 91–92, 148–151; Johanson 1987; Ruane 1994; Stites 1978, p. 167). Kirpichnikova's experiences as a school director as well as participant in this public debate illustrated how women were involved in this shift in institutions and attitudes. Under pressure from reformers, even the conservative Ministry of Education recognized the advantages of expanded access to schools. By 1911, girls made up an estimated one-third of all elementary pupils and almost one-half in urban elementary schools. Yet, only one of every three school age girls attended school regularly and most dropped out before the third grade (Eklof 1986, p. 311). While some elementary schools, especially smaller village schools, educated boys and girls together, most urban schools and almost all secondary schools were separated by gender (Gorchakov and Komarov 1927, p. 45). Throughout this last decade of the Imperial era, coeducation remained a topic of intense public discussion, with lessons drawn from Russian schools and foreign models (see Dunstan 1997a, pp. 383–386).

As Kirpichnikova recognized, the central question "anyone picking up this book will look for" was about relations between boys and girls (Kirpichnikova 1914, p. 83). Kirpichnikova's answer was simple and direct: "I think all of us teachers working in coeducational schools, without arguing among ourselves, will answer the same: relations between boys and girls, between young men and young women, are comradely and ordinary, like those relations that occur among children growing up in the same family" (Kirpichnikova 1914, p. 83). At the same time, Kirpichnikova defended her coeducational school from fears—or charges—that allowing boys and girls to remain in proximity to each other would promote undesirable behaviors: "the flirting that seems most frightening of all is not a common part of coeducation and never takes place" (Kirpichnikova 1914, p. 83). In the end, Kirpichnikova reported, the goal of a coeducational school was to develop "the conviction that simple and comradely relations were possible between people of different sexes," which in turn depended on the commitment, as cited above, to see in each new person "not only a man or a woman, but also a human being" (Kirpichnikova 1914, p. 88).

These concerns about gender differences were also evident in Kirpichnikova's analysis of the evolution in pupils' attitudes. While conceding that the youngest boys, those enrolled in the first grade, "are in the majority of cases confirmed opponents of coeducation," Kirpichnikova dismissed the reasons for this opposition by claiming that they were unhappy just because they could not wear uniforms with caps, insignia, and silver buttons, unlike their peers who attended boys' gymnasia. Kirpichnikova expressed more concern, however, about boys who were teased for attending a "girlie school." Referring to these "blows to the self-esteem of young men," Kirpichnikova approved of those who responded by "speaking in defense of their school and even the idea of coeducation." More ominously, Kirpichnikova reported, more aggressive efforts to defend their school led to "ugly forms," as young boys backed up their claims of "not turning into old women" by exaggerating the number of fights. In this situation, "boys searched everywhere and all over for chances to fight (and found them, despite the most vigilant surveillance) while girls, feeling that they were not wanted, began to keep themselves apart, playing and talking on their own." The aggressive tendencies of boys were thus a problem because their desire to live up to perceived gender roles threatened the broader goal of building a coherent, stable, and cooperative community involving both sexes.

As pupils became accustomed to the coeducational environment, Kirpichnikova argued, they began to display "the consciousness of the commonality of interests" (*soznanie obshchnosti interesov*) defined as the goal of coeducation. Discord among boys diminished and girls were more likely to be included, even in games played in the courtyard. Classes did not divide as sharply in half, as pupils developed more friendly relations with each other and learned to work together, "without distinctions" (*bez razlichiia*). Yet, even as Kirpichnikova argued for the positive effects of coeducation on the community of pupils, she also defined gender differences that persisted in this context. Dedication to their class, concern about the reputation of their school, and a sense of personal responsibility developed earlier among girls than among boys. When teachers needed to figure out some incident, they were more likely to get "objective" statements from girls than from boys.

During breaks, most girls spent the time "walking in twos and threes, conversing among themselves or talking with some of the teachers." Boys, on the other hand, "value any chance to move and run around." These differences in collective behavior were explained in terms of psychological development through the sixth grade level: "Girls already have a need to express themselves and conversation brings them pleasure; boys have not yet developed to this point" (Kirpichnikova 1914, pp. 84–86).

Among older children, by contrast, both boys and girls have the same need to express themselves, and seek opportunities in evening gatherings and excursions, where pupils "talk incessantly" about their activities, companions, expectations, and sympathies. As these differences diminish, the possibilities for community expand: "At no time in groups of just boys or just girls can there be such lively, substantial, and sincere conversations as among the talented pupils of a coeducational school." In the most senior grades, Kirpichnikova reported, the friendship between boys and girls was "so ordinary that neither teachers nor parents are alarmed." While these friendships occasionally developed into a stronger sense of affection (although the dominant tendency was for boys to be friends with boys and girls with girls), even here Kirpichnikova sought to defend the superior virtues of her school:

> We are not afraid of affection between children of different sexes: coeducation, which offers children the opportunity for natural, almost family-like, interaction with each other, is a guarantee that this affection will not develop into abnormal forms of coquetry, courting, and so-called flirtation, in short, into those lamentable and common-place phenomena which can be seen everyday on the streets when pupils of boys' and girls' gymnasia go to their homes.

Arguing that schools needed to prevent the spread of such "vulgar relations," Kirpichnikova invoked "moral reasons" to extend her claim to argue that coeducation was especially necessary in provincial cities and small towns, where so few cultural influences were present (Kirpichnikova 1914, pp. 86–88).

Yet even the academic realm involved notable gender differences in pupils' attitudes, conduct, and relations, as in this description: "Girls conscientiously and thoroughly prepare all assigned lessons; boys begin to avoid subjects for which they do not have a preference." These differences seem readily apparent, and even accepted, according to Kirpichnikova: "If a boy has to ask to be guided quickly through or to have explained an unprepared lesson, it is best to direct this kind of request to a girl, who will have prepared thoroughly and will readily offer assistance." Girls also began to think and worry about examinations, yet it is only when boys begin to show the same concern that "the whole class comes together" and pupils "offer assistance to each other without any difference in gender." In this context, as stated above, boys explained math and physics problems to girls, while girls offered assistance with languages "that the boys must now take more seriously" (Kirpichnikova 1914, pp. 86–87). From this partisan perspective, the differences in the behavior of boys and girls were expected to diminish as pupils recognized on their own the advantages of combing these strengths. Rather than confronting these differences, Kirpichnikova seemed to argue, coeducational schools were better off accepting their persistence with full confidence that over time they would diminish to the point of insignificance.

Kirpichnikova addressed the question of coeducation from a clearly defined liberal feminist perspective (Stites 1978, pp. 191–232). Her opposition to the authoritarian rules and arbitrary practices of the Tsarist regime, her emphasis on the rights of individuals in society, her fear of discord and violence, her belief in a process of gradual and peaceful change, and most of all her commitment to a state-directed educational system that promoted harmony even as it recognized differences represented core beliefs of the Russian liberal movement (Rosenberg 1974, pp. 12–17). Kirpichnikova's arguments for coeducation thus reflected this broader campaign for the development of a public sphere supportive of and guided by an educated elite with clearly recognized forms of professional expertise. Her statement that in the matter of coeducation, "we have no past . . . we have only a limited present and a great future," certainly reflected more generally the progressive views of liberal intellectuals on the eve of World War I (Kirpichnikova 1914, pp. 77–78). Her justification of the school's parent organization as "completely legal," her descriptions of pupils as "independent members of a small society" (*samostoiatel'nye chleny malen'kogo obshchestva*), her promise that coeducation would defend children from "a wave of bureaucratic conventionalism" (*volny kazenshchiny*), and above all her criticism of state interference in running the school provide further evidence of these liberal beliefs (Kirpichnikova 1914, pp. 78–82, 85, 88). Even as she encouraged children to express their own opinions, listen to the views of others, and on this basis develop a binding consensus, Kirpichnikova acknowledged the complex balance involved in this relationship: "Work on your own is not prohibited, but a rare individualist can stand against the common stream, which is diverse and noisy" (Kirpichnikova 1914, p. 82).

This article thus demonstrates the limits of liberal arguments for gender equity. Kirpichnikova anticipated a social revolution that would begin with small changes in behavior, values, and interactions, such as having boys and girls play and learn together at school. Like her liberal allies in prerevolutionary Russia, Kirpichnikova saw such "small deeds" as preparing the ground for Russia to become a society that recognized the rights of individuals, fostered participation in civic organizations, and was governed by elected representatives in a constitutional state. Yet, this vision of the liberal future left unresolved the central tension observed in Kirpichnikova's school among persistent differences in behavior, attitudes, and expectations, on the one hand, and the normative expectation that all members—male and female—of this society would share the same values, on the other. Anticipating the debates of American liberal feminists more than 70 years later, Kirpichnikova believed that boys and girls should be both treated as equals and celebrated for their differences. The subsequent history of coeducation in

Russia, as discussed in the next section, demonstrated that Kirpichnikova's goal of mass coeducation would be realized, but in ways very different from those she had advocated.

More generally, however, her position was reflective of the dilemma of those who believed that pointing out a more "normal" form of behavior would result in certain desired changes in personal, social, or political behavior as a whole. As a revolutionary strategy, the pursuit of gender equity illustrated the fundamental tension in the liberal project, which depends on individuals deciding for themselves to follow the apparently "common-sensical" path laid out by their self-appointed leaders. Such an approach tends to underestimate, however, the power of deeply embedded structures of authority, and thus is frustrated by the seemingly irrational refusal, or inability, of people to act as expected. As is true throughout significant portions of the world today, a comparison to be discussed in the final section of this chapter, the pursuit of gender equity in Imperial Russia illustrated both the great contributions of and the inherent limitations to the liberal approach to large-scale transformations.

COEDUCATION AS A REVOLUTIONARY OBJECTIVE

Kirpichnikova's goal of coeducational schooling was soon achieved in Russia, but not by a government sharing her liberal principles. Mass coeducation followed the 1917 revolution, which overthrew the autocratic Tsarist state and established the world's first communist regime. Driven by Marxist commitment to "emancipating" women and nationalist ambition to overcome "backwardness," the new Soviet regime immediately promised "free and compulsory" education "for all children of both sexes" (*Narodnoe obrazovanie* 1973, pp. 18–19). Building upon the significant advances in the enrollment of girls in mass schools in the last decades of the Imperial era, Soviet educators assumed that coeducational schools would be an essential step toward full equality for women. Yet the goal of providing a full elementary and secondary education to all Soviet children greatly exceeded the resources invested in school development. By the late 1920s, a decade after the revolutionary government proclaimed the goal of educational equality, girls made up just one-third of pupils in Soviet schools.

Soviet educators remained committed, however, to the belief that only a common education for boys and girls would lead to socialism, as in this statement by A. Gorchakov and M. Komarov, two proponents of revolutionary coeducation:

> The economy of the new Soviet society demands, above all, coeducation. In an industrialized context, where men and women work jointly and do the same kind of labor and when it is difficult to draw any line separating men's work from women's work, then it is senseless and useless to talk about separate education . . . The coeducation of boys and girls—future fathers and mothers—from earliest childhood guarantees to them the possibility in their time—at a more mature

age—of standing on the same rung of the cultural staircase. Such a collective will be invincible, such a collective will achieve ever greater victories on all fronts. (Gorchakov and Komarov 1927, p. 44)

Echoing statements made before the revolution by liberals like Kirpichnikova, deputy commissar of education M. Epshtein declared that Soviet coeducation had resulted in "more normal relations" *(k bolee normal'nym vzaiumootnosheniem)* between boys and girls. Looking beyond the school, for example, Epshtein also argued that coeducation would transform attitudes regarding the equality of adult men and women: "a child who learns from the first years of organized and rational life to look at men and women in the same way will exert a known influence on those disgraceful attitudes that persist regarding women, especially in the village" ("Desiat' " 1928, p. 4). Other Soviet proponents declared that a coeducational school was the best place to develop comradeship, cooperation, collectivism, and mutual respect among pupils (Gorchakov and Komarov 1927, p. 44).

Even as coeducation was praised in principle for its transformative possibilities, however, observations of boys and girls at school confirmed that embedded patterns—and perceptions—of gender segregation were still at work. At the same time that he praised coeducation for "making relations between male and female pupils significantly more healthy," for example, Epshtein complained that school groups were still organized along separate lines, so that boys and girls were usually seated on different sides of the room. To deal with these patterns, Epshtein urged that schools not limit themselves to combining boys and girls in the same classes, but should "assume a more active approach to teaching about correct relations between the sexes" ("Desiat' " 1928, p. 4). Revolutionary pedagogues of the 1920s were caught, however, between their dissatisfaction with persistent examples of gender inequities in schools, on the one hand, and their own acceptance of certain seemingly immutable and even desirable gender differences, on the other. In 1927, for example, Soviet educators Gorchakov and Komarov promised that "coeducation would impart to male pupils the softness that is sometimes not available to members of the male gender, and impart elements of persistence and develop will-power among members of the female gender" (Gorchakov and Komarov 1927, p. 44).

The most significant steps toward gender equity came not from the programs and principles of educators, however, but from massive expansion in enrollments in primary and secondary schools. Following the Central Committee's decree on achieving universal education, announced in summer 1930, dramatic increases in enrollment were accomplished, in part through deliberate, sustained, and often coercive use of state power. The "complete enrollment" of girls was defined as one of the goals of the universal education campaign. By 1936, the overall proportion of female pupils in Soviet schools increased to 47 percent and total enrollment of girls more than doubled to 12 million (*Zhenshchina* 1936, pp. 107–108). By 1940, the total number of girls enrolled in elementary and secondary schools had increased to approximately 15 million, including 10 million in rural schools. At decade's end, girls made up 48 percent of school pupils, with a slightly higher proportion (50 percent) in urban schools and a slightly lower proportion (47 percent) in rural schools (Dodge 1966, p. 106; *Kul'turnoe stroitel'stvo* 1940, p. 246). Not surprisingly, this apparent achievement of educational parity figured heavily in Soviet propaganda accounts. The so-called Stalin Constitution approved in 1936 guaranteed both the right to education and the equal rights of Soviet citizens to all men and women (Krupskaia 1938, pp. 162–166; "Vseobshchee" 1937, pp. 3–4).

But these assertions of gender equity concealed as much as they revealed. During the decade after 1917, as described above, Soviet educators had looked to schools to revolutionize practices and thus achieve gender equity. Yet, these revolutionary objectives had their inherent limits. Were the dramatic gains in girls' enrollment matched by equally significant changes in achievement, behavior, and beliefs? What was the relationship between a revolutionary change in *access*—the number and ratio of girls and boys in school—and the *experience* of being male or female at school—not just in terms of the formal curriculum, but also the "informal curriculum" that is often so crucial in shaping gender identities among pupils?[4] While these questions went virtually ignored in Soviet educational discourse of the 1930s, any effort to understand the tension between parity in access and equity in practice needs to ask precisely about how they shaped the complex relationship between pedagogy and revolution. To explore these questions, this chapter returns to the 1939 article by Savich, cited in the introduction, to explore the implications of the "retreat" from revolution on the goal of gender equity.

Coeducation and the End of the Revolution

By the late 1930s, Soviet coeducation had become both an accomplished fact in schools and a striking silence in public discourse. Yet, the title of Savich's article, "Difficult Questions about Coeducation in the School," provides some sense of the contested relationship between the goal of gender equity and the actual practices of Soviet schools. Even as he echoed the revolutionary rhetoric of Soviet discourse by calling for a determined struggle against "legacies of the past and survivals of the old ways," Savich also appealed to educators to develop "a clear and correct understanding of the essence of coeducation" and "a rational use of methods appropriate to having boys and girls in the same school building" (Savich 1939, p. 24). By closely examining Savich's explanation, with its embedded combinations of justification and critique, this chapter concludes that while the revolutionary strategy of achieving gender equity through the schools may have succeeded on the levels of access and enrollment, coeducation failed to bring a genuine transformation of power relations in society, precisely because the pedagogy of gender equity was far less transformative in either intentions or outcomes.

In sharp contrast to Kirpichnikova's position in opposition to the policies of the government and prevailing views of society, Savich's advocacy of coeducation was consistent with government policy, school practices, and seemingly the prevailing assumptions of educators, parents, and the general public. Whereas Kirpichnikova used specific examples drawn from her school to make promises based on an idealized future world of coeducation, Savich identified specific interactions and behaviors that illustrate some of the underlying complexities of coeducation as it actually existed in Soviet classrooms. In particular, Savich called attention to the way that assumptions about gender differences could be observed and how they should be eliminated:

> Any remarks by pupils about boys being more capable or smarter and any similar thoughts by individual teachers must be eliminated from the school as incorrect and harmful. On the contrary, every teacher should demonstrate that girls are working, are mastering the material, and are learning what is being studied in the school just as the boys are. In a class where somehow there has already developed among both male and female pupils a prejudiced belief that girls are weaker, it is necessary to make every effort to help girls ascend to the same level, because any backwardness among girls is usually caused by the incorrect approach taken by the teacher or the sense of depression felt by girls who are "outdone" by boys making use of their greater fluency and undue familiarity. (Savich 1939, p. 28)

Coeducation was thus expected to have transformative implications by first identifying persistent causes of gender equality (such as the belief that girls were inherently less capable) and then becoming the means to eliminate these causes (by exposing and repudiating these assumptions and then taking steps to ensure that girls could achieve at the same level). From this perspective, even in a coeducational school, attention to issues of gender equity brought the potential for a revolutionary transformation in pedagogy.

Yet even as he advocated in favor of coeducation, Savich echoed observations made by Kirpichnikova 20 years earlier by lamenting a persistent pattern of gender separation among the equal number of boys and girls in Moscow schools. As evidence, Savich cited the tendency of boys and girls to sit separately in the classroom and play separately at recess, and generally the lack of "any sense of comradeship and friendship between boys and girls" (Savich 1939, p. 24). These complaints were followed by recommendations of specific measures that could be taken by teachers, such as ensuring the boys and girls were mixed together during field trips, making arrangements so that "a good girl pupil offers assistance to a boy comrade and, on the contrary, a good boy pupil offers assistance to a girl pupil who was lagging behind," or assigning pupils even at the secondary level to sit together in mixed-gender desks. Yet, Savich clearly preferred to follow a gradual, incremental, and thus almost invisible mode of change: "But all this needs to be done without calling attention to why it is being done, so that the combining together happens on its own and so that pupils come to see coeducational work as normal and natural . . . I would like to see classes in which boy pupils sit next to girl pupils without any outside constraints but also without frivolity" (Savich 1939, pp. 28–29).

While recommending these slight, but ultimately quite significant changes in pedagogy, Savich also warned of possible dangers when boys and girls did overcome their separatist tendencies and began to interact more freely with each other:

> Boy pupils are not becoming markedly more gentle and restrained under the influence of their girl-comrades, while girl pupils often exhibit the kind of crudeness and harshness that are not, of course, appropriate for the idea of the new person, the new woman. Boys' carelessness in dress, free and easy way in speech and posture, characteristic lack of discipline, and poor organization are transferring to girls. (Savich 1939, p. 24)

Reasserting that coeducation required more than simply placing boys and girls in the same classroom, Savich warned that crude and undisciplined boys might lead girls in their school to become "unbalanced tomboys who are even more difficult for teachers to straighten out than the most mischievous and naughty boy." Yet the more explicit examples of harmful behavior at the school level all reflected the misbehavior of boys, some of whom harassed girls as they passed through the narrow spaces left in the hallways or peeked into the half-opened doors of girls' bathrooms during recess (Savich 1939, p. 29).

Even as he cautioned against the harmful effect of this mixing of genders, however, Savich also invoked gender differences as a way to discipline pupils in a coeducational school:

> After having seen a male pupil place on the table in front of a female pupil some kind of obscene drawing or a note with cynical language or commit some kind of crude or vulgar joke, I would—right there in front of his classmates, in the presence of girls, or sometimes after pulling him aside if he was particularly naughty—ask him if he would say these words or display such a picture to his mother or sister or what he would do if another boy had, in his presence, treated his sister with such rudeness. The naughty pupil loses his provocative and aggressive manner, begins to think, and occasionally becomes more restrained. (Savich 1939, p. 30)

If such measures do not work, however, Savich recommended "a sharp rebuke." But girls could also be used for disciplinary purposes:

> I always enlist girls to participate in the struggle against the dissoluteness of boys, and I am not afraid to appeal to their feeling of womanly virtue and to their modesty. This struggle depends on the formation of a nucleus of pupils whose support can be used to re-educate the entire class. (Savich 1939, p. 30)

Whereas Kirpichnikova in 1913 had seen coeducation as a way to challenge established customs and if necessary to destablilize accustomed beliefs, Savich saw such traditional cultural norms as "womanly virtue" and girls' "modesty" as effective instruments to achieve the end of maintaining classroom order.

Savich's discussion of coeducation thus assumed a fundamentally ambivalent tone that contrasted sharply to both the advocate's position taken by Kirpichnikova and the prevailing authoritarian, even dictatorial, structure of Stalinist discourse:

> It seems to me that there is not enough clarity in our views on the relationship between boys and girls and in our understanding of what we refer to as communist attitudes toward women. We are somehow afraid to talk about this topic. On the one hand, we are silent because it all seems to clear: we have established full equality (*polnoe ravenstvo*) between men and women in public and political life, and any feelings that result from differences between men and women we perceive as being super-materialist and, afraid of being rebuked for idealism, we are willing to declare unhealthy behavior by individuals to be a healthy result of nature . . . On the other hand, we are afraid that somehow devoting our attention to relations between boys and girls which are based more on affection than friendship will violate some kind of innocent state of the child's being, thus forgetting that these beings long ago passed the age of childhood . . . Setting aside any debate about immutable and eternal feminine origins and leaving to the psychologists the in-depth and detailed study of the psychological traits of pupils, we can clearly see by looking at our children that the combined attendance of boys and girls at school under proper supervision is having a positive affect on one and the other. There is no justification for failing to recognize the positive meaning of this reciprocal influence. (Savich 1939, p. 26)

Even as he acknowledged the limitations, inconsistencies, and tensions in the practices of Soviet coeducational schools, Savich nevertheless concluded with a reaffirmation of how gender equity in schools would promote not only women's emancipation, but also proper communist relations more generally in society:

> Teaching communist relations between boys and girls means the development of a consciousness of complete real equality of the sexes (*soznanie polnogo deistvitel'nogo ravenstva polov*), the creation of a belief that a man and a woman are first of all comrades in the great struggle for socialist development and the bright future of all humanity, the development of respect and consideration for sexual feelings, and, finally, the cultivation of correct and serious understanding of feelings of sexual love. (Savich 1939, pp. 27–28)

Defining the teaching of "propaganda about communist relations in the family, new communist attitudes toward women, and elements of a new morality" as the highest purpose of the school meant, as Savich concluded, that male and female teachers needed to model ideals of gender roles and relations (Savich 1939, p. 32).

Savich's article illustrates the limits of gender equity as a revolutionary strategy in the Soviet context. While reaffirming official claims to full civic and legal equality between men and women, the actual experiences of boys and girls described in this article reveal persistent inequalities. Girls were held to a different standard of behavior that included tighter restrictions on movement and voice. While the formal curriculum may have eliminated grade differences, the "hidden lessons" of teacher attention, the patterns of student interaction, and especially the evaluation of normative behavior testify to persistent inequalities in the experience of schooling. Savich's approach resembled a "deficit model" of pursuing equity, in which girls were evaluated in terms of their relationship to "male" norms and expectations (AAUW 1998, p. 3). This article thus suggests that after more than twenty years, the revolution had not yet achieved the stated goal of gender equity. In fact, the ambivalence and uncertainty of Savich's language testified to the considerable confusion regarding the meaning of gender equity in schools. Perhaps most importantly, the fact that this was apparently the only article in almost a decade to deal explicitly with the education of boys and girls provides the best evidence of how this revolutionary transformation excluded any real consideration of the persistent tension between gender and education.

Conclusion: Gender Equity in a Post-Revolutionary Era

In June 1943, 30 years after Kirpichnikova's advocacy of complete coeducation and four years after Savich's more ambivalent defense of existing conditions, Stalin's government abruptly announced the elimination of coeducation in urban schools. This policy change was justified in terms of the presumed influence of gender on classroom order, academic achievement, and communist upbringing. In 1943, Soviet educators blamed coeducational instruction for overlooking the "particular physical development of boys and girls." Separate schools, by contrast, promised to train male and female pupils for "essential" roles as soldiers and mothers. Some educators went further by recommending that girls' schools reduce technical examples in science lessons, address "everyday" themes in literature lessons, and show concern for "emotionality" (Dunstan 1997b, pp. 170–176; Timofeev 1945, pp. 6–13; Tsuzmer 1943, p. 2).[5] Rather than deal with the ambivalent relationship between equality and difference, Soviet educational authorities apparently decided that separate schools could provide the best education while also fulfilling the Communist Party's political objectives of training boys and girls to become disciplined, patriotic, and obedient defenders of the fatherland and mothers of future citizens.

In the three decades after Kirpichnikova promised that coeducation would "assist the natural development" of boys and girls without "eliminating the difference between the sexes," similar assumptions and objectives continued to shape attitudes and practices in the Soviet Union. This chapter thus confirms the argument made by political scientist G. W. Lapidus more generally about Soviet policy and sexual equality, when she asserted that revolutionary change has meant "not a total rupture with the past, but a partial assimilation and even reintegration of pre-revolutionary attitudes and patterns of behavior that are not merely 'bourgeois remnants' destined to evaporate in the course of future developments but defining features of a distinctive political culture" (Lapidus 1977, p. 117). Soviet educators never resolved questions about the extent to which school policies dedicated to equality should challenge the biases embedded in a particular society, and the extent to which educators should accept the "inevitability" of difference in the name of raising standards of achievement or shaping patterns of conduct. By leaving these questions unresolved, Soviet educators acknowledged the limits of gender equity as a revolutionary strategy. Because gender equity in schools seemed to be neither a clear goal to pursue nor an obvious obstacle to be eliminated, it occupied an ambiguous position in the revolutionary project. While the broader principle of gender equity in education was widely accepted, the actual implementation of this principle revealed the limits of revolutionary change at a practical level.

The question of gender equity in schools was brought back into focus on a global scale in the late 1990s when the Taliban regime in Afghanistan imposed sharp restrictions on the education of girls (Bearak 2000; Clark 2000; Mazurkewich 2000; Pourzand 1999, pp. 78–81). As the United States and its allies intervened militarily in Afghanistan following the September 11, 2001 attacks, these restrictions against girls' schooling were repeatedly invoked as justification for overthrowing a repressive regime (Baker 2001; Mann 2001; Rosenberg 2002; U.S. Department of State 2001). The dramatic increase in the number of Afghan girls attending school in the year that followed was widely interpreted (at least in the West) as a major step toward the emancipation of women, toward equal educational achievement of all, and toward a more fair and just society (Constable 2002, 2003; Dominus 2002; Lacayo 2001). Yet a deeper understanding of gender equity in revolutionary contexts, as illustrated by the materials in this chapter, would suggest a more cautious evaluation of the challenges facing girls not just in Afghanistan, but throughout the world, where girls make up approximately two-thirds of the more than 130 million children not enrolled in school (Sperling 2002).The articles by Kirpichnikova and Savich suggest two different approaches to a shared goal of coeducation: the former appealed to individualist notions of equity and civic values of consensus, while the latter depended on the mobilizing policies of an interventionist state. In the Russian/Soviet context, the former position resulted in shifts in public opinion in favor of coeducation while the latter strategy produced rapid and extensive increases in girls' enrollment. But neither resolved more basic questions about whether educational equity could coexist with gender differences.

This tension remains unresolved in current discourse about persistent girls' schooling around the world. In this era of postrevolutionary change, however, many governments and even more nongovernmental organizations have coupled promises of greater access to education with an effort to address fundamental questions of balancing equity and difference. In a 2002 UNESCO report, a survey of the Gender Parity Index in different regions of the developing world includes categories such as access, enrollment, dropouts, grade repetition, and women in teaching. Yet the broader relationship between gender and schooling that is evident in, but never openly acknowledged in the Russian/Soviet context, is dealt with more directly in the summary report:

> Examining education through a gender lens reveals not only the pervasiveness and the depth of gender disparity in schooling, but also how gender-oriented initiatives can become effective in a relatively short span of time . . . The analysis of gender in education shows that gender equality is a broad societal issue, and it is plain that promoting gender equality in education is part of, and must be accompanied by measures to promote gender equality throughout society . . . Paying attention to the plight of girls in terms of gender disparity also helps reveal areas and levels of education, regions, and countries, where boys are at a disadvantage. In other words, a gender lens allows decision-makers at all levels—politicians, officials, and parents—to see the circumstances, needs, and potential of both girls and boys more clearly. The path to . . . gender equality in education may be better built, not with gender-neutral approaches that flatten the learning environment, but with gender-fair measures that recognize and respond appropriately to the differences between girls and boys (and indeed among other subsectors of the population such as ethnic minorities and persons with disabilities). In this way, a gendered approach to education can bring benefits for all students. (UNESCO 2002, pp. 78–79)

A reading of the Russian and Soviet examples suggests that in the absence of such an explicit recognition of this tension, even the most "common-sense" appeals or the most dramatic large-scale transformations may have relatively little impact on the daily interactions that shape the experience of schooling. In Afghanistan, one year after girls returned to schools, a series of denunciations, threats, and attacks directed at female pupils and women educators have revealed once again how a "revolution" at the political level may have only a limited effect on everyday practices and attidues in and around the school (Constable 2003; Coursen-Neff and Sifton 2003; Harding 2002; Rohde 2002). For those committed to fundamental changes in educational and gender equity, identifying and addressing the need for change at this level must become central to this revolutionary agenda.

Notes

Research for this chapter was supported by a Summer Humanities Stipend from the Department of Interdisciplinary Studies and research funds from the Department of History, Virginia Tech.

1. Savich's article is interesting not so much because of the identity of the author (a Moscow school director and educational department inspector who published a handful of articles on administration, instruction, and writing), but because it seems to have been virtually the only published article specifically about coeducation in the 1930s. The two-volume "pedagogical bibliography" of Soviet books, articles, and chapters includes almost 30,000 entries for 1931 to 1940, but only this one article is listed under the heading "coeducation" (Andreeva et al. 1973).

2. For an even more recent shift in the approach of this extremely influential organization, see AAUW (2001) as well as the highly critical response in Sommers (2000).
3. In general, the secondary literature on women in the revolutionary and Soviet era devotes relatively little attention to schools, even though these institutions provided common experiences for multiple generations. The main exception is the work of Gail Warshofsky Lapidus, who integrates the study of education and women with broader issues of social change and political culture in Communist systems. See Lapidus (1976, 1978a, 1978b). For Soviet coeducation, see also Dunstan (1997), Holmes (1991, pp. 99, 134), Stites (1978, p. 397). My own work has attempted to look beyond the terms of the "gender-less" discourse in order to explore the construction of women teachers' identities, to place Soviet disourse on gender and education in a broader comparative context, and to connect girls' educational access to underlying political campaigns and social transformations (see Ewing 1997, 2002b, c, d, 2005).
4. For the influence of "hidden lessons," see Sadker and Sadker (1994, pp. 1–14).
5. In July 1954, eleven years later and one year after Stalin's death, the government just as suddenly reinstated coeducation in all schools. The Soviet "experiment" with gender segregation is examined in Ewing (2002b).

REFERENCES

American Association for University Women (AAUW). 1992. *How Schools Shortchange Girls. A Study of Major Findings on Girls and Education*. Washington: AAUW Educational Foundation.

———. 1998. *Gender Gaps. Where Schools Still Fail Our Children*. Washington: AAUW Educational Foundation.

———. 2001. *Beyond the "Gender Wars." A Conversation about Girls, Boys, and Education*. Washington: AAUW Educational Foundation.

Andreeva, E. P., N. A. Rut, and N. V. Starikov, eds. 1973. *Pedagogicheskaia bibliografiia* vols. 2–3 Moscow: Prosveshchenie.

Baker, P. 2001. "Teaching Boys—and Girls—Another Notion of Jihad." *Washington Post* October 2.

Bearak, B. 2000. "Afghanistan's Girls Fight to Read and Write." *The New York Times* March 9.

Bil'shai, V. 1956. *Reshenie zhenskogo voprosa v SSSR*. Moscow: Prosveshchenie.

Chirkov, P. M. 1978. *Reshenie zhenskogo voprosa v SSSR (1917–1937 gg.)* Moscow: Prosveshchenie.

Clark, K. 2000. "Afghanistan's Bleak Education Record." *BBC News* April 27.

Clements, B. E. 1994. *Daughters of Revolution. A History of Women in the U.S.S.R.* Arlington Heights: Harlan Davidson.

Constable, P. 2002. "Afghan Pupils Thrilled to Go Back to School." *Washington Post* March 4.

———. 2003. "Attacks Beset Afghan Girls' Schools." *Washington Post* September 8.

Coursen-Neff, Z. and J. Sifton. 2003. "Falling Back to Taliban Ways with Women." *International Herald Tribune* January 21.

"Desiat' let sovmestnogo obucheniia." 1928. *Uchitel'skaia gazeta* June 22.

Dodge, N. T. 1966. *Women in the Soviet Economy*. Baltimore: Johns Hopkins University Press.

Dominus, S. 2002. "Shabana is Late for School." *New York Times Magazine* September 29.

Dunstan, J. 1997a. "Coeducation and Revolution: Responses to Mixed Schooling in Early Twentieth Century Russia." *History of Education* vol. 26, no. 4, pp. 383–386.

———. 1997b. *Soviet Schooling in the Second World War*. Birmingham: Macmillan.

Edmondson, L. 1984. *Feminism in Russia, 1900–1917*. Stanford: Stanford Univetsity Press.

Eklof, B. 1986. *Russian Peasant Schools. Officialdom, Village Culture, and Popular Pedagogy, 1861–1914*. Berkeley: University of California Press.

Ewing, E. T. 1997. "Silences and Strategies: Soviet Women Teachers and Stalinist Culture in the 1930s." *East/West Education* vol. 18, no. 1, pp. 24–54.

———. 2002a. *The Teachers of Stalinism. Policy, Practice, and Power in Soviet Schools of the 1930s*. New York: Peter Lang Publishing.

———. 2002b. "Asserting the 'Particular Needs' of Boys and Girls: Single-Sex Schools in the Soviet Union, 1943–1954." Unpublished paper.

———. 2002c. "Personal Acts with Public Meanings: Suicide by Soviet Women Teachers in the Stalin Era." *Gender & History* vol. 14, no. 1, pp. 117–137.

———. 2002d. "Schooling Against Patriarchy in the 'Non-Russian' Regions: Coeducation in Soviet Schools, 1917 to 1943." Unpublished paper.

———. 2005. "A Stalinist Celebrity Teacher: Gender, Professional, and Political Identities in Soviet Culture of the 1930s." *Journal of Women's History* vol. 16, no. 4, pp. 92–118.

Fitzpatrick, S. 1994. *The Russian Revolution*. New York: Oxford University Press.

Gorchakov, A. and M. Komarov. 1927. "Sovmestnoe obuchenie." *Shkola i zhizn'* no. 10, pp. 43–47.

Harding, L. 2002. "Afghan Fundamentalists Raid Girls' Schools." *Guardian* November 1.

Heward, C. 1999. "Introduction: The New Discourses of Gender, Education, and Development." In *Gender, Education, and Development. Beyond Access to Empowerment*. Heward and S. Bunwaree, eds. London: Zed Books, pp. 1–14.

Holmes, L. 1991. *The Kremlin and the Schoolhouse. Reforming Education in Soviet Russia, 1917–1931*. Indianapolis: Indiana University Press.

Johanson, C. 1987. *Women's Struggle for Higher Education in Russia 1855–1900*. Montreal: McGill University Press.

Kirpichnikova, E. A. 1914. "Iz' opyta sovmestnoi shkoly." In *Sovmestnoe obrazovanie*. N. M. Sokolov and G. G. Tumin, eds. St. Petersburg: Novoe Vremia, pp. 75–88.

Krupskaia, N. K. 1938. *Zhenshchina strany sovetov—ravnopravnyi grazhdanin*. Moscow: Partizdat.

Kul'turnoe stroitel'stvo v SSSR. 1940. Moscow: Gozisdat.

Lacayo, R. 2001. "About Face: An Inside Look at How Women Fared Under Taliban Oppression and What the Future Holds for Them Now." *Time* December 3.

Lapidus, G. W. 1976. "Socialism and Modernity: Education, Industrialization and Social Change in the U.S.S.R." In *The Dynamics of Soviet Politics*. P. Cocks et al., eds. Cambridge: Harvard University Press, pp. 195–220.

———. 1977. "Sexual Equality in Soviet Policy: A Developmental Perspective." In *Women in Russia* D. Atkinson, A. Dallin, and Lapidus, eds. Stanford: Stanford University Press.

———. 1978a. *Women in Soviet Society. Equality, Development and Social Change*. Berkeley: University of California Press.

———. 1978b. "Educational Strategies and Cultural Revolution: The Politics of Soviet Development." In *Cultural Revolution in Russia*. S. Fitzpatrick, ed. Bloomington: Indiana University Press, pp. 78–104.

Lenin, V. I. 1945. "The Revolution and Women." In *Collected Works* vol. 23. New York: Progress Publishers, pp. 300–301.

Mann, J. 2001. "Helping Women is Essential to Rebuilding Afghanistan." *Washington Post* November 23.

Mazurkewich, K. 2000. "Bringing Hope—and Homework—to the Girls." *Time* May 29.

Narodnoe obrazovanie v SSSR. Obshcheobrazovatel'naia shkola. Sbornik dokumentov 1917–1973 gg. 1973. Moscow: Pedagogika.

Orenstein, P. 1994 *SchoolGirls. Young Women, Self-Esteem, and the Confidence Gap.* New York: 1994.

Pourzand, N. 1999. "The Problematic of Female Education, Ethnicity, and National Identity in Afghanistan (1920–1999)." *Social Analysis* vol. 43, no. 1, pp. 78–81.

Rohde, R. 2002. "Attacks on Schools for Girls Hint at Lingering Split in Afghanistan." *The New York Times* October 31.

Rosenberg, E. 2002. "Rescuing Women and Children." *The Journal of American History* vol. 89, no. 2, pp. 456–465.

Rosenberg, W. G. 1974. *Liberals in the Russian Revolution.* Princeton: Princeton University Press.

Ruane, C. 1994. *Gender, Class, and the Professionalization of Russian City Teachers, 1860–1914.* Pittsburgh: University of Pittsburgh Press.

Sadker, M. and D. Sadker. 1994. *Failing at Fairness. How Our Schools Shortchange Girls.* New York: Scribner's.

Savich, A. 1939. "Trudnye voprosy sovmestnogo vospitaniia i obucheniia v shkole." *Sredniaia shkola* no. 5, pp. 23–24.

Scott, J. W. 1988. "Deconstructing Equality-versus-Difference: Or, the Uses of Poststructuralist Theory for Feminism." *Feminist Studies* vol. 14, no. 1, pp. 33–54.

Sokolov, N. M. and G. G. Tumin. 1914. "Predislovie." In *Sovmestnoe obrazovanie* p. v.

Sommers, C. H. 2000. *The War Against Boys. How Misguided Feminism is Harming Our Young Men.* New York: Simon and Schuster.

Sperling, G. 2002. "Educate Them All." *Washington Post* April 20.

Stites, R. 1978. *Women's Liberation Movement in Russia. Feminism, Nihilism, and Bolshevism* Princeton: Princeton University Press.

Timofeev, M. 1945. "Sovmestnoe i razdel'noe obuchenie." *Sovetskaia pedagogika* no. 4, pp. 6–13.

Trotsky, L. 1937. *The Revolution Betrayed. What is the Soviet Union and Where is it Going?* New York: Doubleday.

Tsuzmer, M. 1943. "O razdel'nom obuchenii." *Uchitel'skaia gazeta* August 11.

Tucker, R. 1990. *Stalin in Power. The Revolution from Above, 1928–1941.* New York: Norton.

Tyack, D. and E. Hansot. 1992. *Learning Together. A History of Coeducation.* New York: Russel Sage Foundation.

UNESCO. 2002. *Education for All. Is the World on Track?* Paris: UNESCO publications.

U.S. Department of State. 2001. *Report on the Taliban's War Against Women* November 17.

"Vseobshchee obiazatel'noe obuchenie." 1937. *Rabotnitsa* no. 25.

Zhenshchina v SSSR Statisticheskii sbornik. 1936. Moscow: Gosplan.

3

The Limits of
Pedagogical Revolution

Female Schooling and
Women's Roles in Egyptian
Educational Discourse, 1922–52

Barak A. Salmoni

In the fall of 1918, a small number of Egyptian lawyers, intellectuals, and scions of landed gentry formed a delegation, or *wafd*, to discuss with the British high commissioner in Egypt the possibility of native Egyptian attendance at the post–World War I Versailles Conference. Though the *wafd* envisioned ultimate Egyptian independence, more modest immediate goals involved ending the protectorate imposed during the war and rearranging relations with the British. Rather than accede, the high commissioner rejected the *wafd*'s request out of hand, and denied the legitimacy of the delegation members. The latter then proceeded to tour Egypt, acquiring support for what appeared to the British as a dangerous manifestation of mob activism. *Wafd* leaders were then exiled, provoking mass protest, which in turn elicited British repression, committees of inquiry, and the eventual granting of limited independence to a newly sovereign Kingdom of Egypt in 1922. For Egyptian nationalist intellectuals who lived through the strikes, marches, arrests, and Anglo-Egyptian negotiations, these events became known as the 1919 Revolution. In their eyes, Egypt was experiencing—through the efforts of its own people—a true revolution whereby the country was shaking off the lethargy of the Ottoman past in order to reclaim her ancient Pharaonic greatness. The 1922 discovery of the tomb of Tut Ankh Amon symbolically reinforced this conviction. More fundamentally, Egypt's socially engaged intellectuals considered the political changes of 1919–22 the start of a much broader process: a cultural revolution requiring their own leading participation (Gershoni and Jankowski 1986).

This belief in the dual need for sociocultural and political change paralleled another current gathering strength in Egyptian social thought, namely interest in the role of women in a modernizing Egypt seeking political independence and civilizational parity with Europe (Badran 1995; Baron 1994; Beinin 1998; Botman 1998; Van Vleck 1990). Qasim Amin (1863–1908) had written books advocating for "the liberation of the woman" (*tahrir al-mar'a*) and the creation of "the new woman" (*al-mar'a al-jadida*). Women themselves began to agitate on their own behalf in the years 1890–1910, principally through a "women's press," which called for expanded female societal opportunity and rights, in the context of modernization and opposition to the British. The events of 1919 facilitated a confluence between the nationalist struggle against British occupation and efforts to recast the positioning of women within Egyptian society. As early as March 1919, women took to the streets, joining men in anti-British protests and acting on their own initiative. Soon, women too were martyred by British bullets in the nationalist cause, while pre–World War I feminists now combined women's assertion with nationalist action.

Perhaps the best known of these is Huda al-Sha'rawi (1879–1947), who had established the Union of Educated Egyptian Women in 1914, and was married to a senior member of the early *wafd*. During the 1919 revolution, she organized female street protests, and sent letters of protest to international officials. In 1920, she was elected head of the Wafdist Women's Central Committee, itself a notable example of Egyptian women's autonomous political initiative. Through strikes and boycotts, supporting unemployed workers, and raising nationalist awareness among urban and provincial women, the Central Committee helped the *wafd* to reach a broader spectrum of the Egyptian population. In 1923, al-Sha'rawi and other Central Committee members formed the Egyptian Feminist Union (EFU), again demonstrating the organic link they perceived between nationalist revolution and the struggle for Egyptian women's rights. Later in the decade, the celebrated sculpture "The Awakening of Egypt"—featuring a sphinx roused from slumber by an Egyptian woman unveiling her face—symbolically reaffirmed the connection between national rejuvenation and women's liberation.

Sustaining this connection required continuous attention to altering Egyptian social dynamics. The linkage between nationalist uplift and women's activation thus became central to the thinking of those who viewed 1919 as the beginning of an Egyptian transformation. In particular, the country's cultural elites considered education a central instrument in the revolutionary effort, drawing on a recent past of education's political centrality. For example, Sa'd Zaghlul (ca. 1859–1927), the very leader of the *wafd* who became a national icon and head of the party by the same name, had been education minister from 1906 to 1910, ultimately opposing British policies. Likewise, the political firebrand Mustafa Kamil (1874–1908), who introduced a form of mass politics to Egypt through his political party al-Watan (The Homeland), had established short-lived educational journals, funded schools, and cultivated students as anticolonial activists. Representatives from the Islamic reformist wing of the nationalist opposition also had patronized schools. More generally, the expansion, administration, and curricular content of schooling became central to nationalist resistance to British control of Egypt.

Foreshadowing the post-1919 conception associating nationalist renewal, education, and women's issues, female schooling had been integral to educational policy struggles with the British. The latter had relied mostly upon traditional Qur'an memorization schools, which taught rudimentary reading and arithmetic to rural girls. In particular, their imposition of fees on what had previously been free schooling elicited frequent nationalist criticism of girls' (and boys') education, as did the low quality and number of schools. While urban upper-class Egyptians could send children to more substantive primary schools, there were only two such schools for girls by 1910, with curriculum and fees constantly changing. Feminists found the total lack of state-funded secondary girls schools even more galling. Only one Egyptian woman had earned a secondary diploma by the 1920s, after private study and over strenuous British objections. Along with greater educational provision, Egyptians desired more attention to preparing girls for wife- and motherhood, just as writers on education focused on the need for greater instruction in Arabic and Islam.

Thinking about female education against a background of British occupation ultimately drove Egyptians to consider key relationships linking women, the preservation of national authenticity, and the dynamics of modernized male–female interaction allowing women a unique contribution to nationalist renaissance. An eloquent pre-1919 exploration of these relationships emerges from Ahmad Lutfi al-Sayyid (1872–1963), a contemporary of Mustafa Kamil and Sa'd Zaghlul. Known as the "Professor of the Generation," al-Sayyid proved quite influential on younger Egyptians interested in education. One of Egypt's most prominent nationalist intellectuals and publicists, he founded the Umma (Nation) Party before World War I, was an early *wafd* member in 1919–21, and went on to lead an opposition party in the mid-1920s, later serving as education minister (1928–29) (Ali 1974, 1993). In articles penned for the Umma Party-affiliated newspaper *al-Jarida*, which he edited from 1908 to 1914, Lutfi al-Sayyid repeatedly discussed the nationalist importance of education in empowering Egyptians, asserting that "if our goal is independence . . . it will progress with the advance of the country's education" (al-Sayyid 1912/1937a, pp. 327, 325). While colonialism had ingrained negative traits into Egyptians,

> . . . the consensus of the scientists is that inherited traits, however well-rooted in the people, can be treated through education. . . . Thus if there is a hereditary weakness in the population, an ethical strength obtained through education will come up alongside of it, challenge it, and then wipe out its traces. (al-Sayyid 1912/1945, p. 5)

In short, "education is the ladder of progress and the repository of hopes for national happiness" (al-Sayyid 1912/1937b, p. 329).

Beyond these general assertions, Lutfi al-Sayyid's views on *female* education are more striking as a bridge to attitudes after 1922. Beginning from the dual premise of the need for women to perform their traditional homemaking roles and their new status as equal partners of modern men and guarantors of national strength, Lutfi al-Sayyid—a friend and admirer of Qasim Amin—was adamant that women required an education equal in scope to that of males. In his view, Egyptians of the early 1900s treated their daughters abhorrently, by caring only for the latters' physical attractiveness for the sake of a good marriage partner. Yet, "without an education to purify her from her surroundings' impurities," and bereft of "an instruction to sharpen and enlighten her mind," no Egyptian young woman could be "worthy of her husband's companionship." Men and women were therefore often betrothed without attention to "likeness of the spouses in terms of education and their understanding of instruction," dooming marriages to failure or a merely superficial coexistence (al-Sayyid 1909/1937a, pp. 114–116).

In the opinion of Lutfi al-Sayyid and others, existing Egyptian attitudes to female education were entirely outmoded, since men who benefited from modern, European-style education understood that "comradery based on the enjoyment of companionship for its own sake can rarely be found among those differing in education." The primary purpose of female education was thus domestic bliss (*al-sa'ada al-'a'iliya*), and such bliss was "always built upon concord and similarity between the spouses." Female education, though, was more than simply a matter of a properly functioning home: "domestic bliss is the cornerstone of the happiness of the nation, and the parent who works for this happiness by educating his daughter serves his nation in the most sublime service possible." Speaking to men, al-Sayyid thus portrayed improved female education as essential to both familial harmony and national progress. Indeed, as "the nation is not composed of mere individuals, but of families," women's education was *politically* important, as only it could ensure "the solid family, which deserves to be the unit in the nation's formation, fit for the highest degree of national happiness" (al-Sayyid 1909/1937b, pp. 128, 131).

Though a male with secularist leanings, Lufi al-Sayyid's sentiments were shared by contemporary feminist educators who emphasized the centrality of girls' education to strong nationalist families, as well as by Islamic reformists who criticized European-style schools and called for more culturally authentic girls' schooling, also in order to craft good mothers (*al-Manar* 1908–09; Hashim 1911). Broadly speaking, by the beginning of Egypt's "liberal experiment" in parliamentary democracy (al-Sayyid-Marsot 1977), a lasting conceptual continuum linked nationalism, social regeneration, education, and women. Education was assumed to be a guarantor of national uplift. Yet, the Egyptian nation equaled the sum of its families. Since women were the bulwark of familial bliss, girls' education was an inescapable condition of Egypt's betterment. As wives, mothers, sisters, and daughters, educated females would become the domestic pillar of a strong Egypt. This contribution entitled women to a higher societal valuation, just as schooling was meant to create "likenesses" with educated men, to whom females would be equal. For the sake of motherhood, however, women still required an educational substance different from that of males, implicitly limiting the breadth of their societal opportunities.

Prior to 1919, calls for better education and equal social esteem thus assumed differential curricula and unique roles.

Between 1922 and 1952, belief in the necessity of women's education emerged most strongly among practicing Egyptian teachers, educational administrators, and pedagogical scholars, who collectively constituted a pedagogical community deploying educational discourse to articulate particular visions of sociopolitical change. As "nationalist educator intellectuals" aware of global educational trends (Smith 1990, p. 66), Egypt's pedagogues devoted intense interest to the relationship between women's status and national modernization through education, all along affirming schooling's capacity to change the social order, and recalling the contribution of teachers to the 1919 Revolution (Zaki 1948). This chapter explores this evolving discussion of the justifications, benefits, and modalities of female education in Egypt by investigating vocational teachers' periodicals, academic pedagogical journals, memoirs, and the broader cultural press in Egypt. Broadly, Egyptian educators conceived of women's schooling through a number of connected questions indicating continuities with preindependence thinking: Why should young Egyptian women receive more education than in the past? What should be the roles and comportment of women in a modernizing Egyptian national society? Were women equal to men and entitled to share the latters' space outside of the home? What was the sociopolitical purpose and end result of women's education? Finally, given attitudes toward women's roles and female education's purpose, which curricular elements were of particular relevance to girls?

This investigation of the Egyptian pedagogical community's ideational world reveals a conceptual tension reminiscent of pre–World War I years. While a broad consensus considered increased women's education and an elevation of females' esteem a *sine qua non* of nationalist uplift and modernization, support of wider occupational and societal opportunities for females never seriously shook a conviction in the primacy of motherhood and domestic occupational orientation as women's functions. While hoping to liberate women into a status more equal to men and provide them with opportunities befitting their integral role in nationalist regeneration, Egyptian educators persisted in viewing women as inherently different from men, with unique contributions to make to society. Finally, along with belying the middle- and upper-middle-class origins of pedagogues writing about women, Egyptian educational discourse conveyed a sense that women were not to be liberated for their own sake, but to improve Egyptian society through creating the proper moral and domestic atmosphere for males. Those engaged in programmatic Egyptian educational discourse about women therefore bore a self-consciously revolutionary-modernizing aspiration, along with distinctly nonrevolutionary attitudes. As such, this chapter signals the broader phenomenon of persistent conceptual ambiguities in elite projects of social change. Indeed, reference to educational approaches in the larger Middle East, Asia, and Eastern Europe at the end of this chapter will demonstrate that Egyptians after 1922 were asking the same questions—and were characterized by the same ambiguities—as were their counterparts generally in developing countries, who also sought to modernize while protecting cultural authenticity. Since education is intended to preserve social values *and* generate new sociopolitical realities (de Castell et al. 1989; Olson 1980), the Egyptian experience suggests that pedagogy encourages revolution, only to limit the extent of revolutionary change itself.

ADVOCATING AND JUSTIFYING GIRLS' EDUCATION AFTER 1922

As suggested by this chapter's introductory comments, during the Constitutional Monarchy years (1923–52), Egypt witnessed "the most dynamic and varied women's movement of any Middle Eastern country," represented by several prominent individuals as well as the Egyptian Feminist Union (Keddie 2002, p. 562). Egyptian pedagogical literature dealing with female youth and girls' schooling from the 1920s until the Nasserist coup of 1952 was thus replete with appeals for increased and better female education. Writing in 1922 after a tour of schools in Britain, the lawyer and social critic Abadir Hakim claimed it would be a "moral shortcoming" if Egyptians "neglected the matter of education of the woman for fanciful reasons, and undermine her rights" (Hakim 1922, p. 30). Women educators associated with the EFU made quite innovative use of tradition in explaining the need for more girls' education. Speaking three years after al-Hakim, the female teacher Ihsan al-Qusi looked back to the heroic period of early Islam to assert that the Qur'an commended girls' education, and that 'A'isha and Zaynab, the Prophet's wife and granddaughter respectively, had been learned. Sa'iza al-Nabarawi, EFU leader and protégée of Huda Sha'rawi, took the same approach in 1933. Alternatively, feminist pedagogues linked up with the current of Pharaonicism, then a growing theme in Egyptian nationalism. Drawing a parallel between educated daughters of ancient Egypt's royal house and the cultural–political accomplishments of that era, such advocates held that only with an enlightened female population could Egypt recapture her rightful glory (Badran 1995; Gershoni and Jankowski 1986).

Rather than advocating female education for the sake of women themselves, the majority of pedagogical writers emulated Lutfi al-Sayyid by emphasizing the needs of men and the Egyptian nation as a whole. Girls' education thus went beyond activating one of the two components of human society. Egyptian educators were so convinced of the woman's "tremendous influence on the man," that only female education could prevent "the remaining half's backwardness and decay" (Hakim 1922, p. 31). A 1934 article targeting an audience of elementary school teachers and parents went into further detail regarding the nature of women's impact on man. According to the rural elementary teacher al-Sayyid Bahnsawi, a woman interacted with a man in three capacities exclusively: as mother, daughter (or sister), and as wife. As matron, a woman was *the* ethical and emotional exemplar for her son, such that "the principles which she plants in his mind develop until they become more powerful than his particular emotions." Through moral training and advice, and by providing a proper example, a mother could "condition his habits and develop his instinctive principles." Indeed, for this male advocate of girls' education, George Washington had been inspired by his mother's love of freedom, just as Napoleon Bonaparte was emboldened by his mother's refusal to flee the field of battle. In carrying out her mission, a mother was aided by a unique perceptiveness: "the woman has power to explore the depths of a person and his hidden weaknesses, and the male cannot keep pace with her in this." A mother could in effect totally understand her son's inclinations, needs, and defects, and was ideally positioned to guide him positively (Bahnsawi 1934, pp. 61–62).

According to Bahnsawi, the female also positively shaped a man's life as a daughter. Since she was less able than a male child to contribute to the family's financial welfare, a daughter's material needs forced the father to increase his labors outside of the home. Thereby "zeal is renewed within his soul, and the spirit of initiative rises within him." Along with influencing a man's work ethic, a daughter's comportment as an "angel of morals and virtue" (*mala'ikat adab wa fada'il*) impressed upon all male members of the family the need to desist from rough and uncouth behavior. As a wife, a woman had the most immediately apparent influence on a man: "she possesses the halter of the man and makes him happy or distresses him, and kills him or makes him live according to her knowledge." A proper wife—such as Josephine for Napoleon or Khadija for Muhammad—would raise her husband "to the highest degree of glory and power" (Bahnsawi 1934, pp. 61–62). Ultimately, a woman conditioned the environment of a man at all stages of his life, as she cultured, prepared, monitored, and even controlled him. Understood more restrictively, women possessed no freestanding existence permitting them to focus on their own needs, but were characterized by their male-oriented supportive-conditioning function.

Demonstrating continuity with pre–World War I attitudes, this portrayal of female–male interaction suggests that the most agreed-upon reason for female education was training for proper wife- and motherhood. Writing in 1922, Hakim affirmed that the mission of female schooling was to "help her [a girl] to perform her familial and social duties" (*wajibatha al-'a'iliya wa ijtima'iya*) by ingraining proper emotions and actively explaining to her how to raise a family. In terms of the details, women's education was to "teach her to make the home truly clean, arranged beautifully through praiseworthy economy, and [teach her] how to treat servants, conduct herself towards the people of the house, make her husband enjoy family life, and attract the hearts of her children" (Hakim 1922, p. 131). The primary objective of schooling for both masses and elites was thus education for motherhood. Over the next decade, this emphasis on motherhood remained dominant. More than twenty years later, Muhammad al-'Ashmawi, who had served as education minister 1939–40, spoke of education in terms of readying the girl for her domestic tasks: "for if we prepared the mother, we have prepared the child and the youth and the man. Because she is his mother and sister and wife and

teacher and inspirer" (Ashmawi 1946, p. 301). In short, these Egyptian pedagogical writers desired "an education to help the woman perform her true
duty for which nature has formed her" (Hakim 1922, p. 131).

Yet, according to a female teacher whose sentiments recalled those of Lutfi
al-Sayyid in 1909, because nations were comprised of families, female education assumed a *political–national* character. Since she transformed the home
into "the school of the mother" in which the (male) child learned all the most
fundamental virtues while his mother stripped from him his defects, it was
through a woman that a boy could become "an intelligent man benefiting the
members of his nation and race." According to this woman pedagogue, neglect of girls' education created ignorant mothers who raised young men to be
"an enormous burden on their people and nation . . . because of this kind of
mother the nation will retrogress." It was therefore incumbent upon
Egyptians to awaken to "the service, benefit, and favor of the girl for the
nation, for the truth is that she is the nation" (Hasiba 1934, p. 49).

Just as thinkers from Lutfi al-Sayyid on had pointed to the consequences
of poor female instruction, criticism of government educational policies
became a central vehicle to emphasize the tremendous nationalist importance
of girls' education. After Egyptian independence in 1922, the EFU continued to pressure the government to open girls' secondary schools. In 1925,
the state opened one such school in the Shubra suburb of Cairo, almost four
decades after the first male state secondary school. Further secondary schools
were opened over the next decade, but the pace remained hostage to changes
in government, as did funding. Only in 1929 were females able to obtain the
secondary school diploma, essential for certain kinds of employment. Another
issue of concern to feminist educators was the curriculum encountered by
girls. Though we shall return to this in more detail below, many criticized the
government for insufficient attention to civic, nationalist, and historical educational substance. In the minds of feminist educators, such inadequacies
imperiled girls' potential as nationalist matrons (Badran 1995, p. 143).

Though not explicitly endorsing the EFU's positions, influential male educators implicitly supported their goals. Muhammad Baha' al-Din Barakat,
twice education minister (1930, 1937–38), was highly critical of the state
of female schooling. Because of quantitative and qualitative inadequacies in
government girls' schools, many Egyptian young women, and particularly
those from wealthy families, attended foreign and non-Muslim schools. This
produced "an ethical-national crisis resulting in the most dangerous consequences for the country." In foreign schools, Egyptian girls failed to learn
their nation's history or geography sufficiently, in effect departing from an
Egyptian environment during their formative years. As a result, when a girl
assumed her sociopolitically integral role as a "wife or mother, she was devoid
of the national character, and she educated her children in a way that could
not leave them with any understanding of the Egyptian nation, because she
does not feel that [patriotic] yearning," which develops as a youth. Over the
generations, this deficiency in women's character would engender "an agitation in family life, a disturbance in social customs, and a kind of barrier between

the old and new generation." In contrast, education needed to prepare girls to instill in their own children "love of the nation and national awe" (Barakat 1937, p. 606).

Looking at the prescriptive aspect of Minister Barakat's statements, two elements emerge. First, he indeed felt the basis of women's education to be preparation for domestic life as wives and mothers. Second, he considered that mission an integral contribution to the intergenerational maintenance of national customs, morals, and aspirations. In sum, male and female Egyptian advocates of girls' education justified it mostly in nationalist–domestic terms, so women could ensure that (male) children grew up to think and act as proper Egyptians. Indeed, women's education was repeatedly presented as benefiting *boys*, and as facilitating contributions women made to the functions of the *male* gender. For these thinkers who considered themselves in the conceptual vanguard of Egyptian pedagogy and educational policy formation, improved education was due to women not as an inherent right, but so they could be more effective vessels for the transmission—mostly to males—of "Egyptianness."

THE SCHOOLED *SAYYIDA*
BETWEEN DOMESTIC AND PUBLIC DOMAINS

Though agreeing on the need for improved, more widely diffused, and culturally authentic girls' schooling, Egyptian educators demonstrated a persistent ambiguity regarding the proper function of the educated and modernized Egyptian woman. Not all pedagogical thinkers felt the woman's role was solely domestic. One male writer looked forward to unmarried Egyptian women's increased mixing with men outside of the home, as it would allow the two sexes to become better acquainted with each other, leading to happier marriages based upon love. Further, this mixing would gradually raise society to a level permitting "the sharing (*ishtirak*) of all members in work and the battlefield of life." The author thus hoped women would no longer be prohibited from positions (*waza'if*) outside of the home, though he did not specify which kinds (al-Zahawi 1924, p. 49).

In 1930, Education Minister 'Ali Mahir, who had taken steps to increase the number of women's secondary schools, cautiously supported a broader female presence in the public domain. Advocating greater freedoms for women, he asserted that rather than unrestrained liberty, he envisioned "life where the scope widens for women to use their gifts not only in the general duty (*wazifa 'amma*)—and that is educating families and managing households, given this duty's honor and pride—but also in other affairs which suit her or which circumstances permit her" (Mahir 1930, p. 1040). Mahir thus evinced a guarded, hesitant approach. While prepared in principle to allow for women's societal involvement outside of the home, he did not call for it outright, seeming to prefer that women focus on their "general duty."

Some educators supported a more fully "liberated" status for women. One of these was Munira Thabit (1906–67), a woman educated in state schools and foreign secondary schools, who had gone on to affiliate with the Wafd party and campaign for women's rights, prior to obtaining a law degree in France in the 1920s—a first for Egyptian women. She advocated "total, categorical equality (*musawa mutlaqa*) with men politically, culturally, socially, and materially; this requires that woman equal man in primary, secondary, and higher education." She also insisted on the right for women to vote and engage in political activity on an equal footing with men, and saw "no harm in opening the doors of *some* governmental posts to women." Yet, rather than striving for widespread female government employment, Thabit asserted that women should "first plunge into private [sector] affairs and positions" (Thabit 1925, p. 42). Muhammad Husayn Haykal, the respected litterateur, education minister (1930s–1940s), and protégé of Lutfi al-Sayyid, also celebrated the increased public presence of women. Thanks to Qasim Amin's writings and the growth of female schooling, "we see the woman participating with the man today in various fields of life, . . . and we see respected ladies (*Sayyidat*) undertaking multifarious social reforms" (Haykal 1924, p. 28).

Thabit and Haykal were elite, politically engaged intellectuals. Perhaps even more significantly, certain practicing teachers were just as glad to see women move out of the home. A woman teacher from Mit Ghamr revealed that while "not too long ago her [a female's] leaving home and walking in the street uncovered was considered an unforgivable crime," she could now be seen studying in colleges with male students, even surpassing them on exams. Indeed, several educators took great pride in female students' superiority to males in exam results, publishing comparisons in professional journals. Because of these new opportunities and accomplishments, great strides had been taken in liberation of the woman, such that "we see now that she has shared with the man in all of his affairs and been his close associate in most of them" (Ali 1934, p. 65). As if to provide evidence for this assertion, another teacher from a rural school enumerated several examples of Egyptian women performing "manly" roles. These included the pilot Lutfiya Nadiya, the lawyer and professor Na'ima al-'Ayyubi, the physical education expert Munira Sabri, in addition to other public-sphere women. Along with showing Egyptian women equaling their male counterparts, these examples demonstrated to this author that the women's movement in Egypt in no way lacked for the accomplishments of sister movements in the West. In this respect then, not only was women's societal activation perceived as integral to *national* advancement, but as well, women's accomplishments demonstrated Egypt's *civilizational* status. Like-minded teachers hoped for the diffusion of educational opportunity to the lowest rural levels, so that "the female villagers in the countryside, [who] remain ignorant and do not know a thing" could also become conscious, civilized citizens assisting in the larger goal of establishing Egypt's credentials as a full partner in global, that is, European, modernity (Khalifa 1934, p. 66).

Emerging from the pen of a male author, these last words indicate that such sentiments were not limited to the teaching corps' females. Perhaps more important, these articles were written by teachers in small towns or rural areas. As discussed earlier, urban-based, upper-middle-class pedagogues either focused their concerns mostly on the education and societal status of girls from their own socioeconomic backgrounds, or implicitly viewed matters through such a filter. By contrast, teachers in the educational trenches of the Egyptian countryside concentrated on the masses of Egyptian girls. Upper-middle-class voices always exerted more influence on the tone and focus of discourse, and this bias will emerge again with reference to evolving curricular attitudes. Still, among rural educators there were teachers just as stridently committed not only to better girls' schooling, but also to a comprehensive broadening of females' societal opportunities. This commitment also characterized Muslim Brotherhood-affiliated teachers, often sent to rural areas as a sort of political exile. Though emphasizing the importance of female veiling and subservience to men, they too wanted more girls schools so as to hinder missionary and foreign schools' access to Muslim youth—a concern shared by Minister Barakat in the urban, elite milieu (Fitri 1934, pp. 47–48; Qasim 1933, pp. 26–27). The few schools established on Brotherhood initiative included vocational training, implicitly legitimizing activities in addition to motherhood. Indeed, though deserving further research, one may suggest that teachers in rural schools—themselves of non-elite roots—were not as committed as their urban counterparts to a modern domesticity constraining women's public roles. Of course, we may also note that both groups of educational practitioners and writers viewed it as their—and the state's—task to school modern Egyptian youth out of the ignorance presumed to animate informal, familial education.

By far the most unequivocal advocate of women's activity in the public sphere was Asma' Fahmi, assistant dean of the Women's Pedagogy School. Writing in 1947, she began with the assertion that "one of the first purposes of modern education" was to grant the girl "a total social preparation." This meant that the era of providing knowledge "for a narrow, limited occupation" such as motherhood had come to a close. The school had now to abolish those "ancient traditions" in the name of which families "make [girls] lose their self-confidence and pass on to them fear" of the world around them. Rather, educational experiences needed to "prepare the young girl for her new societal duty," which was to include full political participation. In particular, Fahmi envisioned "social mobilization for girls" through compulsory national service. Not military in nature, it would serve "the realization of social and humane goals, since the duty of the woman is humane before everything else." Through this kind of national service, women would gain "several of the skills and mores necessary for proficiency in public services, and their guidance in the correct national direction." Tapping into a theme then popular in Egyptian civics and ethics curricula, Asma' Fahmi declared that living a life of order and joint effort for the sake of a common, higher goal "is as necessary for a young woman as it is for a young man" (Fahmi 1947).[1] She thus envisioned a thoroughly public role for the modern Egyptian woman, where the latter would participate with men in the bettering of society. Still, her activities were not simply to match those of men;

instead, she was encouraged to find her own (social and humane) field in which to make an equal national contribution.

For many, women's proper social and humane sphere comprised educational work in the schools. As early as 1925, women accounted for nearly a quarter of new graduates of state teachers schools (Wizarat al-Ma'arif 1926, pp. 12, 16, 19). In contrast to earlier reservations, the new professional consensus by the 1930s was that even female teachers who married should be allowed to continue in the schools, as long as they met their domestic responsibilities. Amir Boktor, the American-trained professor of pedagogy at the American University in Cairo who edited Egypt's most prestigious professional journal of education, represented this view best. In his view, permitting married women to continue in schools would avert the loss to the teaching profession of experienced female teachers. In this regard, new appliances had made it less time-consuming for women to care for the family, thus freeing them to work as educators. Furthermore, "life is not complete except through marriage, and if we were to deprive married women [of educational work], we would thus deprive education of the complete woman." Additionally, no bright, active woman would remain content solely as a housewife, and "it is a mistake to deprive them of work in the most noble profession, which is the profession of education" (Buqtur 1935, p. 137). Notably, in defending married female teachers, Boktor still took pains to demonstrate they could still fulfill their primary role as a parent and wife. Of course, by approving of women's teaching, educators continued to advocate the traditional female role of conditioning the world of others, though now through rearing nationalist youth in schools.

Notwithstanding voices in support of a public though nurturing role for women, by 1952 a broad pedagogical front advocating sustained functions for Egyptian females outside the home had yet to emerge. In 1941, the male head of the department of pedagogy in the Women's School of Education stated that while he accepted in principle that women could pursue higher education, it was "more appropriate to male youth" because it did not really serve the "preparation of the girl for her natural duty." Here, he cited Herbert Spencer's view that education must prepare people for a complete life; according to the author, "the perfect life for women only exists in their performance of their wifely and motherly duties in the best and most complete manner" (al-Makhzanji 1941, pp. 284, 287). He then went on to report the progress of European schools in providing a domestic-oriented curriculum to female students. Such a rhetorical strategy permitted the author to support a rather traditional position for women in society through reference to supposedly modern European thinkers and methods. Likewise, Minister Muhammad al-'Ashmawi hoped that by the year 2000, Egyptian women would finally recognize that to compete with men in the latters' own arena outside the home "is a matter not benefiting her, not honoring her, and not benefiting the country; it is good for her to cling to her incontestable possession, and to work in her sphere on educating the generation, for that is what will raise her to immortal rank" ('Ashmawi 1943, p. 458).

An official of the Justice Ministry illustrates quite well the overarching hesitancy to liberate women completely from traditional roles, which encapsulated efforts toward male–female educational equality. In an article written in a primary teachers' professional journal only three months before the 1952 Revolution, Mahmud Shawqi repeated the refrain that women were an essential component of society, with a domestic duty only they could perform: "she is the mistress of the house as a wife and mother—and she is the first partner in the family—and she is the financial planner of this institution or mini-state (the family)." On her talents depended the bliss of the entire family and the security of nations. Because of the important function she performed, she was guaranteed significant social rights. Yet, it was more worthy of her to "take a position of patience and waiting as regards different [legal and political] rights," so that those that she gained over time would not conflict with each other.

Shawqi then indicated that just like women, men too enjoyed different privileges based on their social status and circumstances. By not pressing the issue too stridently, he implied, women would ultimately gain complete equality as a group. The author then opined that "equality between man and woman is guaranteed by the nature of creation . . . and I find no objection to the woman having rights equally as does the man in different fields of work." Far more important than any of this, however, was "our need for reform which demands that the woman play a weighty, worthy role in the uplift of the nation." While the Egyptian people as a whole would progress through formal schooling, such education was "accompanied by what every individual learns of ethics and education in his home and family, over which the woman has control." Though different from the man's activities, such a function "is no less grave than any role a man performs" (Shawqi 1952, pp. 6–7).

Thus, in spite of a notable evolution of attitudes to women's roles by 1952 including (at least rhetorical) acceptance of male–female equality, the notion of women undertaking public roles of equal merit to those of men never displaced the still-dominant conception of women as primarily wives and mothers, domestically separate from men. Likewise, even strong supporters of women's advancement, such as the politically engaged litterateur Ahmad Amin, viewed women as essentially different in nature from men. While women balanced and reined in men's excesses, they still mostly facilitated the latters' advances (Amin 1934, pp. 32–35).[2] As this approach suggests, increased "liberation" of females was presented in terms of its benefit to males. And, while pedagogues perceived schooling as granting women higher status and greater skills than in the past, education was not intended to position them to rival men. In this respect, schooling was meant to prepare women to complement men in nationalist terms, but from within the domestic sphere.

Coeducation and a Feminine Curriculum

Beyond general advocacy of women's education and abstract visions of female social roles, pedagogical thinkers and practicing teachers also confronted the quite concrete issues of school dynamics and curriculum. As for the former, were girls to study *with* boys? In certain parts of the Middle East during these years, such as Turkey and the Zionist community of Palestine, coeducation became a policy priority of elites hoping to revolutionize social relations (Arat 1997; Elboim-Dror 2001; Salmoni 2002). Yet, unlike Turkey's Republican regime or socialist Zionists, Egyptian elites did not aspire to *totally* rearrange social conventions. Furthermore, Egyptians had not historically considered coeducation a criterion of female educational equality. Notably, advocates of female Egyptian educational expansion in the 1890–1914 period had not discussed coeducation at length. In Beth Baron's perceptive view, broadening girls' schooling in these years involved "extending . . . segregation into the classroom" (Baron 1994, p. 124). After 1922 however, coeducation[3] began to receive attention among Egyptian pedagogues. Significantly, it was in the context of coeducation that reference was made to other countries' experience, in a manner allowing a closer look at Egypt's cultural associations and aspirations. Further, the benefits of coeducation were often presented as improving a *boy's* character and educational experience. Displaying this tendency, an early article from 1924 pointed to the positive experiences with coeducation in England, Germany, and Norway. The author was impressed that girls and boys both studied and played together. "This education mellows and polishes the ethics of the boys while it also imparts to the young girls firmness in constitution and vigor of body" (*al-Hilal* 1924, p. 865).

Amir Boktor dealt with this topic in much more detail a decade-and-a-half later. There was no question in his mind that coeducation on the elementary/primary and tertiary levels was the only choice in modern civilized societies. Only a small group of "reactionaries" (*fi'a raja'iya*) opposed this kind of mixed education in the supposed name of traditional authenticity, while in reality they sought to prevent female competition with their sons on exams or for employment. Further, other countries emerging rapidly from backwardness also espoused coeducation. Indicating Boktor's view of Egypt's developmental peer group, these countries included China, Japan, the Philippines, India, and Iran, in addition to Turkey. Thus, the only way for Egypt to avail itself fully of the political, economic, and civilizational benefits of modernity was to school its girls and boys together. As for the latter, left on their own they were often insufficiently balanced, reliable, or academically persistent:

> The presence among [boys] of girls [however,] forces them to observe general decorum, choose expressions well, and to eschew clowning around and licentiousness, while being judicious and weighing their words, . . . for [a man] tries not to do in front a woman what he would not do in front of his sister or mother.

Intellectually, "female students generally have a tremendous amount of mental poise, wisdom and discernment," which boys could emulate (Buqtur 1937a, pp. 396–397).

By presenting the advantages of coeducation for male students, Boktor justified the practice in general. This approach may have been tactical, so as to meet the reservation of coeducation's opponents head-on. Still, it matches quite well the general casting of women's instrumental, supportive relationship to men in Egypt during these years. Concluding his article, Boktor considered whether coeducation was possible in the sensitive years of secondary schooling. He noted that in Spain, Iran, and Turkey—those countries perceived as most relevant to Egypt—the general policy did not support coeducation during these adolescent years, though Turkey had recently experimented with it to good end. Finally, schools were coeducational at all levels in the United States as well as several smaller European countries and the Phillippines, therefore commending mixed schooling to Egypt. Thus, in spite of reactionary groups in and out of school, Boktor noted that "it gladdens us that the enlightened class, responsible for the sons of the nation, is comforted by the education of the young girl and her enjoyment of those rights that the young boy has" (Buqtur 1937a, p. 400). Significantly, though Boktor spoke as a member of the urban upper middle class, he lauded coeducational mass schooling in the United States and elsewhere, implying approval of it both in Egypt's rural and urban sectors.

Compared with the topic of coeducation, Egyptian educators were much more concerned with what kind of school subjects were most relevant to girls. In the realm of curriculum, a quite open and non-gender-specific attitude toward female educational substance characterized several pedagogues. The pioneer female student, educator, school patron and pedagogic thinker Nabawiya Musa summed up the view as "equal plus more" (Badran 1995, pp. 144–145).[4] Yet, after appeals for male–female curricular parity during the 1920s–1930s, Egyptian educational thinkers began to view girls as needing a curriculum distinct from that of boys. In 1927, Sa'iza Nabarawi had criticized a prominent Cairo girls school, because "to appease male opinion concerning female education a large portion of the program is set aside for house-keeping, home economy, and cooking." Nabarawi, who campaigned in the 1930s for women's suffrage, desired a curriculum just as full of general subjects as was the case for Egyptian boys. This was particularly true in nationalist terms. The EFU repeatedly appealed to the Education Ministry to include civics material in the girls curriculum to the same degree as was the case for boys schools, in order to "prepare [girls] to be good citizens, and to make women aware of their rights and responsibilities" (Nabarawi 1927, p. 9, 1930, p. 4). By contrast, curricular inequity would leave the door open to foreign schools' socialization of girls in a nonnationalist fashion, thereby guaranteeing that "our daughters will imitate foreign ways" (Badran 1995, p. 146). As Nabarawi's concerns here mirrored those seen above of male education ministers and Muslim Brotherhood-affiliated teachers, her comments highlight the broad consensus on the importance of nationalistically authentic female socialization.

By the 1930s, however, even feminists began to emphasize the uniquely domestic curricular needs of girls. In 1937, Na'ima al-'Ayyubi regretted that the clamor for male–female curricular parity had caused "an ignorance by our girls about the woman's duty towards their children, family, and society." What

Margot Badran has astutely called a "renewed emphasis on women's maternal roles" was reflected as well in professional pedagogues' writings on educational substance (Badran 1995, p. 148). In 1930, Amir Boktor expressed consternation that syllabi in girls schools "follow boys schools letter for letter." In particular, he desired more attention to drawing, music, cooking, needlework, childcare, clothes washing, and other subjects unique to "women's needs." Seven years later, he and his British guests were thoroughly impressed with the Amira Fawziya state secondary girls school in the Bulaq quarter of Cairo, which he felt to be exemplary not only for Egyptian schools, but for European ones as well. A chief reason was its curriculum attuned to what he felt girls needed. Drawing classes focusing on ornamental designs "built upon [girls'] inclination to this branch of fine arts." Visitors to the school also found classes about cooking, sewing, embroidery, and care for small children. Further, extra-curricular activities included visits to poorer families in order to look after their needs and learn a frugal lifestyle. Boktor thus concluded "how good it is for the Education Ministry, the nation, and female students" that such an innovative school existed (Buqtur 1937a, p. 263).

The substance of a curriculum particular to girls thus combined those skills deemed necessary for motherhood with attention to fine arts. M. H. al-Makhzanji, head of pedagogy at the Women's Teachers School, felt home management (*tadbir manzili*) to be so important that the final years of a young woman's education should specialize in it (al-Makhzanji 1941, p. 287). Others believed fine arts (*al-funun al-jamila*) were just as relevant, since they aided moral development and activated one's spiritual inclinations. More concretely, one writer felt that "drawing, painting, and engraving, in this era almost enter into the rank of the necessities." Along with teaching girls good taste, "this practical knowledge, among affairs women can pursue from within [the home], has become a source of earning and even capital from which a materially needy woman can benefit" (Bayhum 1923, pp. 164, 166–167). Finally, skills such as poetry and music merited inclusion in girls' schools, as they were deeply rooted in the Arab past and Arab women had demonstrated their talents in them throughout history.

Even women educators speaking in favor of increased public rights and roles still felt girls needed gender-specific curricula. Asma' Fahmi was no exception. On the one hand she asserted that for women to fulfill their new social duties, they needed to study history, geography and civics, "to direct the attention of the female student to the problems of society and its needs." On the other hand, Fahmi believed women needed to cultivate traits setting them off from men, and allowing them to accomplish their societal goals in a gender-appropriate fashion. These traits could be described as "social graces" or "social elegance" (*al-rashaqa al-ijtima'iya*). Along with physical exercise, girls needed to "be concerned with the ways of conviviality, conversation, and social intercourse, and the arts of entertainment such as music and singing." Further, girls schools needed to devote more time to training their students in "the good selection of colors and apparel, and the methods of beautifying the home, [as well as] grace and elegance in gait and movement." This kind of indoctrination (*tathqif*) was more necessary for girls than for boys, and any school experience excluding it was a "failing, ossified education" (Fahmi 1947, pp. 80–82).

'A'isha 'Abd al-Rahman's (1913–98) perspective on the women's movement in education is illuminating in this regard. Still a child during the 1919 Revolution and not closely affiliated with the Egyptian feminist movement, by 1941 she had obtained a Masters degree in Arabic after winning an award for a book on the Egyptian countryside's problems. In 'Abd al-Rahman's view, by struggling for equal rights and social status, Egyptian women had neglected to set a final goal or define the ultimate essence of a unique womanhood. In this respect, she criticized a supposed lack of curricula sufficiently attuned to women's needs. Lamenting that plans for girls' education, from architecture to syllabi, had simply been borrowed from that of boys, she asserted that this approach "had resulted in a plainly evident mistake in girls' education." Any difference in school environment or subjects for girls was merely formalistic such that "it was desired that she study what a boy studied in terms of math and science and mechanics and chemistry, in spite of the difference of [boys' and girls'] nature and the difference in their roles in life." To the extent that there was any sort of curriculum particular to women, it was defective. Indeed, appliances in secondary schools' home management classes were much more expensive than most girls were likely to find in their future homes. Ultimately, the author claimed Egyptians lacked a clear vision: "Ask them: what is the purpose of girls studying these topics, and what is the goal aimed for in girls schools?" While 'Abd al-Rahman supported women's rights and desired male–female equality, she felt that the movement as a whole, and its educational component in particular, had lost sight of the need for a character unique to Egyptian women, infused with a domestic role and those "social graces" to which Fahmi referred ('Abd al-Rahman 1941a, p. 22, 1941b, pp. 24–25, 1941c, pp. 21–22).

As alluded to above, the socioeconomic perspective of several educational writers informed their discussion of the proper curriculum for females. According to affluent urbanite intellectuals such as Boktor and Fahmi, the topics needed in schools under the rubric of fine arts appear to match the kind of skills appropriate to middle- and upper-class Egyptian women. Indeed, drawing, music, and a fashion sense were considered quite relevant to well-off Egyptian families, and matched what these parents would want to see their girls learn. Of course, as we saw above, certain writers acknowledged that some skills associated with this group could benefit impoverished women as a source of income. Such sentiment may signify recognition of the needs of less-privileged Egyptian young women. Still, most discussion of the quantity and substance of female education during these years implicitly neglected rural schooling and educational tracks for poorer Egyptians, demonstrating an inclination to perpetuate an affluent stratum of women endowed with social refinement and the skills associated with motherhood. In this respect it is instructive to note that the secondary schools founded in the late 1920s and 1930s were built in Cairo, Alexandria, and a few provincial centers, and catered to middle- and upper-class families who could pay fees. Similarly, renewed concern for feminine curricular needs in the 1940s resulted in Feminine Culture Schools for this same social group. Sustained interest in curricular substance for poorer girls was evident only among the small number of technocratic career administrators who had grown up in rural poverty and were known for their trenchant criticisms of all aspects of Egyptian educational policy (Ali 1974; al-Qabbani 1944, 1951; Dayf 1977).

Put somewhat differently, educators such as Asma' Fahmi and 'Abd al-Rahman felt that Egypt's *Sayyidat* needed to be equipped to aid in Egypt's sociopolitical uplift in a way unique to their gender, just as they needed to preserve their characteristic modern bourgeois domesticity, also considered essential to Egyptian development. Such a reenergized concern for motherhood-oriented curricula toward the end of Egypt's "liberal experiment" in multi-party politics implied a certain constancy with views dominant in the early 1900s. Here we may return to Ahmad Lutfi al-Sayyid to grasp the continuities. On the eve of World War I, he had encouraged increased attention to Egyptian women's education in order to cultivate a strong nation based on domestic harmony and the male–female attitudinal resemblances (*mushabahat*) undergirding political–cultural community. Supporting female education in 1950, Lutfi al-Sayyid repeated arguments going back to 1908. Increasing numbers of educated men required that young women receive the kind of schooling permitting them to be appropriate wives. Naturally, ignorant women could not provide such modernized men with "marital happiness." As in 1908, in 1950 Egyptian men and women required those similarities or resemblances (*mushabahat* in 1908; *shibh* in 1950) that "education creates among the schooled . . . especially if the method of education was one [the same]." Thus, only women's education would provide "the familial happiness which is the basis for all other happiness," including that of the nation in general (al-Sayyid 1950, p. 7).

CONCLUSION

From the 1919 Revolution until the 1952 Nasserist coup, practicing teachers, educational administrators, and pedagogical thinkers whose views influenced the teaching corps in Egypt considered improved female schooling and broadened social roles intimately connected to national uplift as a whole. As seen in this chapter, enthusiasm among pedagogues for female schooling tapped into the post-1919 revolutionary tenor, while celebration of Egypt's educational accomplishments in this field allowed pedagogical writers to assert political–cultural membership in a group of European and successfully Europeanizing countries. Further, expanding on a pre–World War I trend, Egyptian pedagogical discourse was characterized by the prominent and continuing participation of women themselves, either as feminist advocates, educational administrators, or teachers. More broadly, sustained discussion of girls schooling and the contributions of educated women to Egypt's modernization reflected educators' firm faith in the pedagogical community's centrality to producing a social order worthy of newly gained political independence.

In maintaining a constant call for better girls' schooling and broadened socioeconomic opportunities, and by beginning to contemplate coeducation, the Egyptian pedagogical community proposed a role for women that was revolutionary in the historical and regional context—though views were often articulated in an instrumentalist framework. Significantly, young Egyptian women who benefited from such an educational discourse later became feminist critics of the social contradictions of Egypt between independence in 1922 and the Nasserist coup in 1952. Such developments demonstrate that elements of the educational approach examined here surpassed general social attitudes (Ahmed 1999; al-Sa'dawi 1999, pp. 80–98). Indeed, while worthy of further inquiry, recent research suggests that these open perspectives toward women in Egypt from the 1920s to the 1950s contrast with the constricted female role envisioned from the 1970s to 1990s, both by Islamists as well as by a state attempting to cope with oppositional religious activism through socially conservative retrenchment (Abu Lughod 1998, pp. 243–269). This remains a complex issue however, requiring nuanced examination; as growing numbers of Islamic activists since the 1990s have focused on nonviolent societal re-Islamization as opposed to political militancy, they have emphasized the importance of Muslim women's education. Likewise, since the infrastructurally inadequate secular state has failed to deliver quality schooling to female (as well as male) citizens, an alternate Islamic sector has begun to provide girls with an education preparing them for properly informed, active Muslim motherhood. In this sense, as distinct from views of professional educators or contemporary feminists, male and female Islamic activists themselves are now participating in a redefinition of girls' knowledge, education, and social roles. This topic should prove quite interesting in the future (Rosefsky-Wickham 2002, pp. 133–134, 162–175).

As mentioned at the outset of this chapter, alongside its potential for social change, education performs a preservative function regarding societal mores. Accordingly, in spite of innovative aspirations regarding women's schooling and social status, educators—including most women involved in pedagogical debate—possessed traditional perceptions about the proper character and activities of females. The latters' access to schooling was justified, in the 1950s as in 1909, in terms of preparation for motherhood, just as those promoting coeducation portrayed it as advantageous to boys. Likewise, initial departures in approaches to women's roles and relationships with men during the 1920s and early 1930s were overshadowed in the late 1930s and 1940s by renewed focus on feminine particularisms. This "regression" may be related to the faltering of Egypt's parliamentary order and the rise of competing Islamicist or quasi-fascist ideologies, though further research is needed to verify this connection.

Educational inquiry beyond pedagogical discourse also demonstrates a reenergized concern for a properly feminine schooling. In the realm of curriculum, from the late 1940s Egypt's Education Ministry established Feminine

Culture Schools and Feminine Vocational Schools, partly in response to desires of women educators. Of impact on a greater number of students, Egyptian secondary curricula in the late 1930s and 1940s began to include a substantial complement of domestic-oriented studies in a track unique to girls, in a sense reducing the latters' exposure—relative to boys—to those general culture courses able to instill the national feeling the EFU had desired (al-Husri 1949, pp. 456–459, 1954, pp. 360–362; Salmoni 2002, pp. 685–712). Likewise, Arabic language texts on the primary and secondary level were often gender-coded, presenting males as boy scouts, athletes, and group leaders, while portraying girls as aspiring mothers and homemakers. In the same vein, history texts criticized Pharaonic-era female cultic or political involvement and celebrated Muslim matrons (al-Din et al. 1939, pp. 3, 9; 'Awamiri et al. 1946, pp. 75–76; Idgar and Ghurbal 1931, pp. 45–48).

Egyptian women's recollections of school-life in the 1940s also highlight the tension between educators' desire for increased girls' modern schooling and their own conservative attitudes to women's roles. For example, Nawal el-Saadawi found that her father, an educational administrator who encouraged her learning, was still much less impressed with her success in school, and much more focused on his son's poor performance. His complaint— "I wish she had been born the boy and he the girl"—shows that Egyptian educators held firmly to the conviction that increased women's access to modernizing schooling should not threaten the beneficial components of traditional feminine morality and submissiveness (and male prominence), so essential to anchoring the nation's uplift in cultural authenticity and social equilibrium (Sa'dawi 1999, pp. 154–155).

Egyptian attitudes therefore reflect the larger Middle Eastern context in which women's modernization was considered integral to nationalist uplift and cultural reform, on condition that "modernized" women continued to focus on domestic roles. For example, in Turkey from 1923 to 1950, girls' education was part of a larger campaign to reform the polity and de-Islamize social conventions. In the new Turkish Republic coeducation was a regime priority, and educational officials celebrated female teachers, doctors, judges, and pilots in the pages of teachers magazines and school texts. Yet, though male–female political equality was legislated and presented as a given of the new populist political order, both educational discourse and curriculum encouraged women to think of themselves as mothers and nurturers of rising (male) generations (Arat 1997, p. 100; Salmoni 2002, pp. 575–584).

Likewise, in Iran from the Constitutional Revolution years (1906–11) forward, women's situatedness in the evolving conception of the nation was, at least in men's eyes, contingent upon their mothering function, even while pedagogical thinkers repeatedly appealed for greater female educational opportunity and curricular breadth. Such sentiments continued into the Reza Pahlavi years (1920s to 1941), and became a major component in the educational effort. Early on, the state opened several new girls' schools, while pedagogical thinkers and curriculum made reference to politically prominent pre-Islamic Iranian women and Islamic-period female poets. Yet, educators' commitment to female political liberation was limited, while "the school of patriotic 'housewifery' " was meant to teach girls that "a mother's role in spawning and nurturing the next generation of Iranian patriots ranked among her sacred duties" (Kashani-Sabet 1999, pp. 187, 189, 194). In French-colonized Lebanon and Syria, male nationalists during these years tended to support education for women *only if* it prepared them to be mothers and housewives, while some Muslim writers discouraged school-based girls' education, fearing female exposure to European and Christian mores (Thompson 2000, pp. 122, 220, 275–276).

The case of the Levant suggests the Egyptian nationalist-pedagogical view was more liberal than that of some of their neighbors, implying a correlation between political regimes and approaches to women's education and status. Here, multiparty Egypt occupies a mid-point between the authoritarian, sociopolitically reformative mono-party Republic of Turkey on the one hand, and the colonized regions of the Levant on the other. These latter lacked political–ideological unity, and faced imperialist challenges to nationalist assertion and cultural authenticity rendering attitudes to women much more defensive. Turning to Iran, though Reza Shah proved no less authoritarian than Turkey's Kemal Atatürk, he lacked total sovereignty, did not nurture visions of social revolution, and possessed a smaller cadre of pedagogical experts. Though leaders in Tehran focused on removing education from the purview of religious clerics, Iranian attitudes to women's schooling and social roles were likely closer to those in Syria and Lebanon than in Egypt. These tentative suggestions indicate the tremendous benefits to be had from future comparative educational inquiry on the topic of gender and schooling.[5]

More broadly, notions regarding the importance of female education—yet for motherhood-directed purposes—appear to have predominated among developing nations in general during these years. In China, expanding educational opportunity for girls was considered a central method to rescue the country from national weakness, through a focus on having the schools create "good wives and wise mothers" (Judge 2001, p. 772). The latter phrase had even been borrowed from Japanese educational modernizers of the 1880s–1930s who felt that only properly moral matrons could mediate youths' exposure to Western enlightenment. Asian educators thus also wished girls' schooling to preserve "the customary virtues of humility, patience, and self discipline, [and] a willingness to sacrifice for the family" (Marshall 1994, pp. 45, 108). As another chapter in this volume makes clear in reference to Eastern Europe, though Soviet education—with which Egyptian pedagogues were peripherally aware—was much more committed to coeducation and equitable educational provision, Russian teachers also presented the former as beneficial to boys, just as by the end of the 1930s "traditional beliefs [such] as 'womanly virtue' and a shared acceptance . . . of female modesty" were seen as vital to classroom order and proper social relations (Ewing 2005).[6] The case of Egypt is thus quite important in demonstrating how educators in developing countries in the first half of the twentieth century explicitly sought to attain modernized gender dynamics that still implicitly reflected male-dominated, traditional conventions. In the conceptualization of working pedagogues—that important segment of the schooling venture that articulates and attempts to implement a national educational vision—society was to be revolutionized through educational advances, but that very educational effort was to preserve mothers', sisters', and daughters' primary patriotic role as domestic and ethical supports to men.

NOTES

1. For resonance with curriculum, see Salmoni (2002, pp. 902–919, 1025–1043).
2. According to this litterateur, sometime politician, and school text editor, men took a broad view of things, formulating general plans to accomplish large tasks. Women examined small details of matters, looking out for practical problems. Thus, while men hatched great schemes in the public sphere, women would troubleshoot them and maintain the health of the home in all its small details, ensuring that men's plans did not fail. While a woman was thus supportive here as well, the male–female relationship resembled a symbiosis.
3. Translated into Arabic as "al-ta'lim al-mushtarak" (joint education), or "al-ta'lim al-mukhtalit" (mixed education).
4. The first girl to sit for state baccalaureate exam in 1907 (she scored in the top third), Musa (1886–1951) began founding girls schools from 1909, becoming headmistress of the Women Teachers Training Schools in Mansura and Alexandria in 1910 and 1915. Though not allowed to attend the embryonic Egyptian University, she lectured at its women's auxiliary from 1909. After a brief period of service as the Education Ministry's chief inspector for girls schools—she was dismissed for criticizing sexual harassment and misconduct—she went on to found several girls schools and write widely on the topic of female education.
5. This is particularly the case for twentieth-century North Africa. The varied impact on gender conceptions of French colonialism, Islamic resistance, local tribalism, and indigenous state centralization in this region has been traced quite adeptly by Mounira Charrad (2001), inviting an examination of strategies for deploying education in connection with gender, particularly in Tunisia, relative to Egypt and other regional countries.
6. For two somewhat skeptical Egyptian views of Soviet education that admit of Russian innovations, see Khaluf (1925) and *al-Tarbiya al-Haditha* (1936).

REFERENCES

'Abd al-Rahman, 'A'isha. 1941a. "Ta'lim al-Banat fi Misr 2." *al-Thaqafa* 3, 151, 17/11/41.

———. 1941b. "Fi Ta'lim al-Banat." *al-Thaqafa* 3:156.

———. 1941c. "Ta'lim al-Banat fi Misr 1." *al-Thaqafa* 3:150.

Abu-Lughod, Lila. 1998. "The Marriage of Feminism and Islamism in Egypt: Selective Repudiation of a Dynamic of Postcolonial Cultural Politics." In *Remaking Women: Feminism and Modernity in the Middle East*. Abu-Lughod, ed. Princeton: Princeton University Press.

Ahmed, Leila. 1999. *A Border Passage: From Cairo to America—A Woman's Journey*. New York: Penguin.

'Ali, Astirah. 1934. "Aydan. . . ." *Sahifat al-Ta'lim al-Ilzami* 7 (March).

'Ali, Sa'id Isma'il. 1974. *Qadaya al-Ta'lim fi 'Ahd al-Ihtilal.* Cairo: 'Alam al-Kutub.
————. 1993. *Nazarat fi al-Fikr at-Tarbawi.* Cairo: Dar Sa'd al-Sabah.
Amin, Ahmad. 1934. "al-Furuq al-'Aqliya Bayna al-Rajul wa-l-Mar'a." *al-Hilal* vol. 43, no. 1 (November).
Arat, Yesim. 1997. "The Project of Modernity and Women in Turkey." In *Rethinking Modernity and National Identity in Turkey.* Sibel Bozdogan and Resat Kasaba, eds. Seattle: University of Washington Press.
al-'Ashmawi, Muhammad. 1943. "Misr fi 'Amm 2000, Hayatuna al-Ijitma'iya: Tanabbu'at Muhammad al-'Ashmawi." *al-Hilal* vol. 51, no. 4 (October).
al-'Awamiri, Ahmad, Ahmad 'Ali 'Abbas, 'Awad Lutfi Ahmad, 'Abbas Hasan. 1946. *al-Mutala'a al-Mukhtara li-l-Madaris al-Ibtida'iya, Juz' III, Sana III.* Cairo: al-Matbaa al-Amiriya.
Badran, Margot. 1995. *Feminists, Islam, and Nation: Gender and the Making of Modern Egypt.* Princeton: Princeton University Press.
Bahnsawi, al-Sayyid. 1934. "Ta'thir al-Mar'a." *Sahifat al-Ta'lim al-Ilzami* vol. 8.
Barakat, Muhammad Baha al-Din. 1937. "Hal Addat Wizarat al-Ma'arif Risalataha?" *al-Hilal* vol. 45, no. 6 (April).
Baron, Beth. 1994. *The Women's Awakening in Egypt.* New Haven: Yale University Press.
Bayhum, Muhammad Jamil. 1923. "al-Mar'a wa-l-Funun al-Jamila." *al-Hilal* vol. 32, no. 1–2 (October–November).
Beinin, Joel. 1998. "Egypt: Society and Economy, 1923–1952." In *The Cambridge History of Egypt.* M. W. Daly, ed. vol. 2. Cambridge: Cambridge University Press.
Botman, Selma. 1998. "The Liberal Age, 1923–1952." In *The Cambridge History of Egypt.*
Buqtur, Amir. 1935. "Zawaj al-Mu'allimat: Hal Yajuz Istimrar al-Mu'allima fi Wazifatiha Ba'd al-Zawaj?" *Majallat al-Tarbiya al-Haditha* vol. 9, no. 2 (December).
————. 1937a. "Madrasa Misriya Raqiya li-l-Banat." *Majallat al-Tarbiya al-Haditha* 10:3 (February)
————. 1937b. "al-Ta'lim al-Mushtarak Bayna al-Jinsayn." *Majallat al-Tarbiya al-Haditha* vol. 10, no. 4 (April).
————. 1930. "Khalaya al-Nahl fi-l-Madaris al-Misriya." *Majallat al-Tarbiya al-Haditha* vol. 3, no. 4 (May).
Charrad, Mounira. 2001. *States and Women's Rights: The Making of Post-Colonial Tunisia, Algeria and Morocco.* Berkeley: University of California Press.
Dayf, 'Abd al-Salam, Shawqi, *Fikr Ismail Mahmud al-Qabbani wa-Athrihi 'ala Tatawwur atl-Tarbiya wa-l-Ta'lim fi Jumhuriyat Misr al-'Arabiya.* MA Thesis: College of Education, University of Manuqiya, 1977.
de Castell, Suzanne, Allan Luke, and Carmen Luke. 1989. "Beyond Criticism: The Authority of the School Textbook." In *Language, Authority and Criticism: Readings on the School Textbook.* de Castell, Luke, and Luke, eds. New York: The Falmer Press.
al-Din, Ibrahim Numayr Sayf, Zaki 'Ali, Ahmad Nagib Hashim. 1939. *Misr fi al-'Usurr al-Qadima.* Cairo: Matba'at al-Ma'arif.
Elboim-Dror, Rachel. 2001. "Israeli Education: Changing Perspectives." *Israel Studies* vol. 6, no. 1.
Ewing, E. Thomas. 2005. "Gender Equity as a Revolutionary Strategy: Coeducation in Russian and Soviet Schools." *Revolution and Pedagogy.*
Fahmi, Asma'. 1947. "al-Taliba al-Misriya Bayna al-Madrasa wa-l-Mujtama'." *al-Hilal* vol. 55, no. 11 (November).

Fitri, Muhammad. 1934. "al-Hijab." *Sahifat al-Ta'lim al-Ilzami* vol. 1, no. 6.

Gershoni, Israel and James Jankowski. 1986. *Egypt, Islam and the Arabs: The Search for Egyptian Nationhood, 1900–1930*. New York: Oxford University Press.

Hakim, Abadir. 1922. *al-Tarbiya al-Akhlaqiya*. Cairo: Matba'at al-Yaqza.

Hashim, Labiba. 1911. *Kitab al-Tarbiya*. Cairo: Fatat al-Sharq.

Hasiba, Zaynab Muhammad. 1934. "al-Mar'a allati Tuhizz al-Mahd bi-Yaminiha Tuhizz al-'Alam Bi-Yasariha." *Sahifat al-Ta'lim al-Ilzami* vol. 1, no. 7.

Haykal, Muhammad Husayn. 1924. "Ba'd Qasim Amin." *al-Hilal* vol. 43, no. 1 (November).

al-Hilal. 1924. "Inqilab fi Ta'lim." 32:8 (May).

al-Husri, Sati'. 1949. *Hawliyat al-Thaqafa al-'Arabiya I*. Cairo: Arab League; Matba'at Lajnat al-Ta'lif wa-l-Tarjama wa-l-Nashr.

———. 1954. *Hawliyat al-Thaqafa al-'Arabiya III*. Cairo: Arab League; Matba'at Lajnat al-Ta'lif wa-l-Tarjama wa-l-Nashr.

Idgar, G., M. Shafiq Ghurbal. 1931. *Kitab al-Ta'rikh al-Qadim, li-Talimidh al-Sana al-Ula al-Thanawiya*. Cairo: Matba'at al-Ma'arif.

Judge, Joan. 2001. "Talent, Virtue, and Nation: Chinese Nationalisms and Female Subjectivities in the Early Twentieth Century." *The American Historical Review* vol. 106, no. 3.

Kashani-Sabet, Firoozeh. 1999. *Frontier Fictions: Shaping the Iranian Nation, 1804–1946*. Princeton: Princeton University Press, 1999.

Keddie, Nikki R. 2002. "Women in the Limelight: Some Recent Books on Middle Eastern Women's History." *International Journal of Middle East Studies.* vol. 34, no. 3.

Khalifa, Husayn Muhammad. 1934. "al-Nahda al-Nisa'iya." *Sahifat al-Ta'lim al-Ilzami* vol. 8 (April).

Khaluf, Mikha'il. 1925. "Ghara'ib al-Ta'lim fi al-Madaris al-Bulshafiya." *al-Hilal* vol. 33, no. 9.

Mahir, 'Ali. 1930. "Nahdat al-Mar'a wa Ta'limuha: al-Fatat al-Misriya fi-l-Madaris al-Ulya." *al-Hilal* vol. 38, no. 9 (July).

al-Makhzanji, Muhammad Husayn. 1941. "Hawl Tarbiyat al-Banat: al-Tarbiya li-l-Umuma." *Majallat al-Tarbiya al-Haditha* vol. 14, no. 4.

al-Manar. 1909. "Bahth fi Khutbat al-'Aqila al-Misriya 'Bahitha bi-l-Badiya.' " *al-Manar* vol. 12, no. 6 (1327).

al-Manar. 1908. "Muqaddimat Mutarjim Kitab Tarbiya al-Istiqlaliya Ba'd al-Basmala." *al-Manar* vol. 11, no. 6 (1326).

Marshall, Byron K. 1994. *Learning to Be Modern: Japanese Political Discourse on Education* Boulder: Westview Press.

Nabarawi, Sa'iza. 1927. "L'enseignement secondaire féminin en Egypte." *L'Egyptienne [al-Misriya]* (November).

———. 1930. "Deux interviews avec le Ministre de l'Instruction Publique." *L'Egyptienne* (July).

Olson, David R. 1980. "On the Language and Authority of School Textbooks." *Journal of Communications* vol. 30, no. 1.

al-Qabbani, Ismail Mahmud. 1944. *Siyasat al-Ta'lim fi Misr*. Cairo: Matba'at Lajnat al-Ta'lif wa-l-Tarjama wa-l-Nashr.

———. 1951. *Dirasat fi Masa'il al-Ta'lim*. Cairo: Maktabat al-Nahda al-Misriya.

Qasim, Muhammad Fathi. 1933. "al-Mar'a Bayna al-Sufur wa-l-Hijab." *Sahifat al-Ta'lim al-Ilzami* vol. 1, no. 4.

Rosefsky-Wickham, Carrie. 2002. *Mobilizing Islam. Religion, Activism, and Political Change in Egypt.* New York: Columbia University Press.

al-Sa'dawi, Nawal. 1999. *Daughter of Isis: The Autobiography of Nawal el-Saadawi.* London: Zed Books.

Salmoni, Barak. 2002. "Pedagogies of Patriotism: Teaching Socio-Political Community in Twentieth-Century Turkish and Egyptian Education." Ph.D. diss, Harvard University.

al-Sayyid, Ahmad Lutfi. 1912/1945. "al-Tarbiya wa-l-Ta'lim." *al-Jarida* 1684 (24/9/12), in Ahmad Lutfi al-Sayyid, *al-Muntakhabat II.* Cairo: Matbaat al-Muqtataf.

———. 1912/1937a. "Ila al-Shabiba 5: Wasa'il al-Istiqlal." *al-Jarida* 1669 (4/9/12), in Ahmad Lutfi al-Sayyid, *al-Muntakhabat I.* Cairo: Dar al-Nashr al-Hadith.

———. 1912/1937b. "Ila al-Shabiba 6: Wasa'il al-Istiqlal, al-Tarbiya wa-l-Ta'lim." *al-Jarida* 1670 (5/9/12), in Lutfi al-Sayyid, *al-Muntakhabat I.*

———. 1909/1937a. "Banatuna." *al-Jarida* 610 (14/3/09), in Lutfi al-Sayyid, *al-Muntakhabat I.*

———. 1909/1937b. "Salah al-'A'ila Salah al-Umma." *al-Jarida* 630 (6/4/09), in Lutfi al-Sayyid, *al-Muntakhabat I.*

———. 1950. "al-Walidun wa-l-Awlad." *al-Hilal* 58:6 (June).

al-Sayyid-Marsot, Afaf Lutfi. 1977. *Egypt's Liberal Experiment, 1922–1936.* Berkeley: University of California Press.

Shawqi, Mahmud. 1952. "Huquq al-Mar'a wa Wajibatuha." *al-Mu'allim al-Awwal* vol. 8 (April).

Smith, Anthony D. 1990. *National Identity* London: Penguin.

al-Tarbiya al-Haditha. 1936. "Hal Taqud Rusiya al-Sufiyatiya al-'Alam fi al-Tarbiya wa-l-Ta'lim?" *Majallat al-Tarbiya al-Haditha* vol. 10, no. 2 (December).

Thabit, Munira. 1925. "Nahdat al-Mar'a al-Misriya: Hadith Ma'a al-Anisa Munira Thabit." *al-Hilal* vol. 34, no. 1 (October).

Thompson, Elizabeth. 2000. *Colonial Citizens: Republican Rights, Paternal Privilege and Gender in French Syria and Lebanon.* New York: Columbia University Press.

Van Vleck, M. Richard. 1990. "British Educational Policy in Egypt Relative to British Imperialism in Egypt, 1882–1922." Ph.D. diss, University of Wisconson, Madison.

Wizarat al-Ma'arif. 1926. *Taqrir Yubayyin Hal al-Ta'lim Alladhi Tatawwaluhu Wizarat al-Ma'arif aw Tushrif 'Alayhi fi Akhr December 1923.* Cairo: al-Matba'a al-Amiriya.

al-Zahawi, Jamil Sidqi. 1924. "al-Mar'a al-Sharqiya." *al-Hilal* vol. 33, no. 1 (October).

Zaki, Ahmad. 1948. "Nasha'tu fi al-Thawra . . . Taliban wa Mudarrisan!" *al-Hilal* vol. 56, no. 10 (October).

4

Pedagogies and Politics of "Culture"

Chiefly Authority, the State, and the Teaching of Cultural Traditions in Ghana

Cati Coe

Wunni panyin a due ampa.	If you don't have an elder, my condolences.
Mmerewatia ne nkwakorawa yi Wɔne nkwadaa na edi afoofi.	These old, old women and men, It is they who stay home with the children.
Wɔyɛ akyerɛkyerɛfoɔ pa. Wɔkyerɛ kasa pa ne mpanyinsɛm	They are good teachers. They teach proper language and elders' matters
Tete mmofra nimdeɛ kwan so.	and educate children to be knowledgeable.

—Poetry recital from Kwawu South, Eastern Regional
Cultural Competition, April 16, 1999
(translation by Afari Amoako and Cati Coe)

In southern Ghana, as this poem, recited as part of a school cultural competition, declaims, it is through elders that one learns forms of knowledge—appropriate speech and elders' matters—that are marked as cultural. The poem hints that a child gains access to these skills through dedicated copresence with elders while staying at home doing housework. Furthermore, the verb *kyerɛ*, translated here as "to teach," means also "to show"; the verb *sua*, commonly translated as "to learn," originally meant, "to imitate."[1] Cultural education by elders in southern Ghana takes place through informal, daily interaction, emphasizing performance and activity through demonstration and imitation, rather than verbal instruction or inner awareness. This is knowledge *how to* rather than *knowledge of.* Thus, two systems of education exist side by side in southern Ghana; both an informal education connected to a gerontocratic social order, in which elders are the custodians of the most prestigious knowledge, and formal schooling linked to the rise of the modern nation-state, the training of experts (teachers), and the creation of a new elite. The history of schooling in Ghana has in some ways been about the relationship between these two realms of a child's experience, with those concerned with schools and the state attempting at various times, sometimes simultaneously, to denigrate and appropriate the power associated with traditional social order and the education associated with it.

This essay explores the interactions and tensions between these two forms of education and authority through the lens of the teaching of Cultural Studies in the region of Akuapem in southern Ghana. The subject of Cultural Studies was introduced as part of a larger educational reform package in 1986, instituted by a government that was considered revolutionary at its beginning but which by the mid-1980s was working with the World Bank and International Monetary Fund (IMF) on a structural adjustment program. Several questions arise from this series of events and policies: why did the government launch the Cultural Studies curriculum and subject? How did it fit or not fit either its populist rhetoric or its neoliberal economic and educational policies? Finally, even if a reform of revolution is said to "fail," what kind of work is the Cultural Studies curriculum nonetheless accomplishing, both ideologically and on a practical, day-to-day level?

The problem of the Ghanaian revolution has long defied categorization. In 1981, Flight Lieutenant J. J. Rawlings and the Provisional National Defence Council (PNDC) took over the government of Ghana through a military coup, proclaimed a "people's revolution," and declared war on the wealthy and external dependency. As economic and environmental hardships deepened in 1982, the PNDC courted international financial institutions, thus paving the way for neoliberal economic reforms as part of IMF mandated structural adjustment programs in 1983 and 1987. Thus, despite the revolutionary rhetoric of the PNDC, which explained the economic crisis as a result of Ghana's dependency on an international economy controlled by multinational industries and transnational banks, its policies focused on managerial factors such as mismanagement and inefficiency and resulted in economic liberalization (Folson 1993; Ninsin 1991). To what extent can this be considered a revolution? Why did the government decide to woo the IMF and World Bank and accept structural adjustment programs? Was this simply a difference between rhetoric and practice? What factors within Ghana, such as a need for the government to gain control over local authorities, shaped the PNDC government's apparent retreat from a revolutionary campaign against neocolonial dependency?

Paul Nugent has described the 1981 revolution as an "ambiguous revolution," in which the PNDC government was unable to sustain its unity of purpose as various factions, including leftist student organizations, trade unions, and technocrats within the government, competed over the trajectory and pace of the revolution (Nugent 1995). As a result of the economic hardships of 1982, the technocrats gained credibility over left-wing factions within the PNDC government, who were increasingly sidelined.

Eboe Hutchful sees Ghana's revolution as essentially an institutional revolution within the state, carried out by a group of disaffected junior military officers, rather than a popular or social revolution. At the same time, he warns, one should not necessarily distinguish between social and institutional revolutions, as hierarchies within the state bureaucracy were linked to larger social inequalities in Ghana (Hutchful 2002). The state is a primary generator of social advancement and inequality, and the Ghanaian elite is created and sustained by its connection to the state, including educational institutions. By the end of the 1970s, "the decay of the state and public institutions had precipitated a profound questioning from within of authority structures, norms, morality, and operating styles" (Hutchful 2002, p. 35). Thus, capturing, reforming, and controlling state institutions, including schools, became the prime objective of the 1981 revolution.

If we see the 1981 revolution's aim as the state itself, the pursuit of neoliberal economic reforms makes sense. Although neoliberalism is associated with the paring back of the state in favor of private enterprise, in Ghana neoliberalism was consistent with the aggrandizement and renewal of the state, albeit with a new form of extractive politics. In part, this was because the Ghanaian team that worked with the economists from the IMF and the World Bank did not share their vision about the increased efficiency and productivity of market forces compared with the state, which the state should imitate (Hutchful 2002). However, neoliberalism continues to see a role for the state to play; the government must actively construct political, legal, and institutional conditions under which the market can exist (Burchell 1996). Under neoliberalism, the state does not simply disappear as markets take over; rather, it plays different roles (Clarke 2002). Furthermore, the orientation to global sources did not end up undermining the state in Ghana, but rather strengthened it, because the Ghanaian leadership used the increased flow of funds to recharge the state (Hutchful 2002). The primary effect of the 1981 revolution was the increased presence and power of the state in the Ghanaian landscape.

We see a similar conjunction between the practice of neoliberalism and a rhetoric of anticolonialism in the PNDC government's educational reforms. The World Bank-sponsored educational reforms of 1986 entailed putting more resources into the first nine years of education—defined as Basic Education—at the expense of secondary and university education, reducing the total number of years of preuniversity schooling, making Basic Education more vocational so that those who did not go on to secondary school could be self-sufficient and self-employed, and introducing school fees that increased with each level of schooling so that, outside of teachers' salaries that were paid by the government, schools could be self-supporting. This reform pushed through some initiatives that had a long history of being proposed—and resisted—in Ghana, such as vocational education and reducing the total number of years of schooling (Foster 1965).

Despite the switch to neoliberal economic and educational policies, the Cultural Studies program and a new Cultural Policy became a site for a rhetoric of cultural authenticity and antineocolonialism. "The Cultural Policy of Ghana" begins with this statement:

> Colonialism and imperialism were both initially installed through instruments of violence and subjugation. To sustain colonialism, however, it was necessary to employ the instrument of cultural imperialism. This consisted of the total denial of our history, the denigration of our system of values and the replacement of our essential religious, social, political and economic structures with structures carefully fashioned to ensure the perpetuation of the subjugation of our people, the nurturing and enhancement of an inferiority complex in our personality and the continuous servicing of both the ego and the material well-being of the colonial metropolis by the colonial structures so established. (National Commission on Culture 1991, p. 1)

The cultural policy aimed to "replace the inherited structures with ones which will redress the imbalance in the psyche of our people caused by the totality of our colonial experience" by placing new emphasis on Ghanaian culture within colonial institutions, like schools. The PNDC government hoped "to ensure the structuring of the total schooling environment from its physical organization to its pattern of interpersonal relationship such that an authentic Ghanaian message as well as pride in the Ghanaian cultural heritage is easily transmitted to the student" (National Commission on Culture 1991, p. 8). However, the educational reforms it instituted concentrated and bounded cultural practices as an academic subject, similar to other subjects such as math, social studies, and science. Thus, the Western model of the school remained; "culture" was inserted as a subject, to be taught and examined like other subjects, within the context of a classroom, notebooks, and a state-written curriculum.

We return to one of the initial questions posed: why did the PNDC government institutionalize the teaching of "culture" in schools? One reason is that as national governments lose more control over their economies to international regulations and agreements, they turn to culture and pedagogy as sites of national concern and authenticity. In many postcolonial states, the legitimacy of the state depends on the provision of social services (Boli and Ramirez 1986). A second reason is that, since independence in 1957, "culture" has been nationalized and secularized by the government, which also appropriates its transmission through school. Like previous national governments in Ghana (Rathbone 2000; Yankah 1985), the PNDC government has attempted to undercut independent chieftaincy, align chiefs to the national state, and use symbols of chiefly authority to legitimate their own (Owusu 1989). State events and celebrations provide occasions for cultural performances, and representatives of the state have attempted to appropriate local cultural festivals in Ghana for state purposes. Thus, cultural activities become ways and opportunities for the state to showcase itself. So, if at a national level, revolutionary talk about culture looks antineocolonialist, at a local level, it expresses the state's appropriation of symbols of local authority. Thus, neoliberal educational reforms are consistent with a nationalist cultural education policy if we see the revolution's project as strengthening the power of the state, with the state becoming more of a presence in people's everyday lives.

However, educational reforms are always a project, entailing certain imaginations and intentions that may never be realized (Thomas 1994). This essay therefore explores how the state's appropriation of forms of cultural knowledge associated with local social structures were actually realized in classrooms, at the level of teaching and learning, focusing on students' and teachers' agency. It particularly assesses the effects on authority, with specific reference to the transformation of cultural knowledge into school knowledge, the limited expertise of teachers in relation to elders and some students, and the limits on the state's ability to appropriate local cultural traditions through cultural competitions. I first lay out the historical and social context of Akuapem before analyzing how culture gets produced in school classrooms and its effect on the status and authority of teachers, elders, and youth.

Studying Akuapem and the Nation: Context and Methodology

The interaction of the traditional kingdom of Akuapem with the state began in the 1890s, as the British colonial administration became more powerful after the defeat of the Asante Empire in 1896, and colonial administrators were heavily involved in mediating chieftaincy disputes. To this day, the modern nation-state coexists and overlays a traditional kingdom in which the 17 towns of Akuapem are organized into a hierarchical arrangement based on military formations, speaking to the importance of warfare in the past (Kwamena-Poh 1973). Although chieftaincy has lost its political power gradually since the beginning of the twentieth century, chiefs retain control over the sale of land and continue to be respected as "custodians of culture" and spiritually powerful. They are not an independent authority set apart from the state. Rather, they are subject to the state's overtures and appropriations. The national government of Ghana has consistently sought to use chieftaincy to cement its local legitimacy, at the same time as chiefs have sought legitimacy by bringing development funds to their regions (Dunn and Robertson 1973). Chieftaincy in Ghana is intensely factional, which the government exploits, supporting one chiefly candidate over another, at the same time as it deplores the fissuring of traditional states—including Akuapem in the 1990s (Gilbert 1997)—as causing instability. Most chiefs in Akuapem are members of the elite and professional class and live in the urban centers of Koforidua and Accra, returning on weekends for festivals and funerals. These are some of the ways that royal families attempt to use their educated members, whose money and influence give them access to the state, to promote the fortunes of their families and town.

The position of chiefs illustrates a point that is also true of the larger population in Akuapem; although people's loyalties and identifications revolve around their towns, rather than the traditional kingdom or the nation, Akuapem is dependent on its connections outside, to both rural and urban areas. People in Akuapem wait for money from relatives working in the cities of Accra, Tema, and Koforidua and are dependent on food from the valleys to the west or from food-producing areas even farther away. It is caught in the middle of the urban–rural divide in Ghana, with the eyes of its youth turned to the cities and the old people remembering the ways of a hunting and farming life. Its architecture testifies to the importance of connections to the outside world: to the Basel missionaries who built their houses and schools in a new style that was widely copied; to the money earned through cocoa, for which Akuapem cocoa-farmers began migrating to the west for land in the 1890s (Hill 1963); and to current migrations to the cities and overseas.

I lived in Akropong, Akuapem for a year (August 1998 to August 1999) and also visited for a five-week period during the summer of 2002. During these research trips, I visited schools in the area and teachers in their homes, observing classroom lessons on "culture" and other subjects, and attending church services and traditional festivals. Ultimately, I focused on two primary schools and one junior-secondary school. As secondary schools began preparing for cultural competitions in February, I accompanied the district cultural studies officer as he gave workshops to teachers, and then picked three secondary schools to follow as they rehearsed for and performed in the competitions. I interviewed judges of the competitions and conducted focus group and individual interviews with student performers after the competitions.

In many ways, the economy of Akuapem is education, beginning with the Basel Mission's establishment of a seminary, which later became a teacher training college, in 1848 and a boys' boarding school in 1867 (Kwamena-Poh 1980). Teachers' salaries are a source of cash for their relatives in Akuapem (Brokensha 1966), but the boarding institutions also support large catering and groundskeeping staffs. Akuapem's educated children are the ones able to succeed in their current migrations to cities and overseas, sending remittances home. Education is both highly valued and in better shape in Akuapem than it is elsewhere in semiurban areas of Ghana. Akuapem is known in Ghana as having a long history of involvement with the twin projects of Christianity and education; both are associated in the minds of many Ghanaians with the loss of "culture." This region is thus an ideal context in which to examine the pedagogies and politics of culture.

MEANINGS OF CULTURE IN AKUAPEM

Several meanings of culture operate in Akuapem, but I focus on the one most pertinent to the discussion of youth and age. A popular view sees culture (or *amammrɛ*) as historical, involving town and family history as well as the ritual practices and wisdom of the ancestors and elders. Most cultural and historical knowledge is called *mpanyinsm* or "elders' matters," and the distribution of this prestigious knowledge attests to the gerontocratic hierarchy operating in Akuapem.

A popular phrase refers to chiefs as the "custodians of culture." The chiefs, aided by the knowledge of their elders, are the ones who perform the rites that maintain good relationships with the spirits and ancestors of the town and family, on behalf of the whole community. Kwame Ampene, a retired teacher of Akan language and music, articulated this most clearly:

> The chiefs are custodians of our culture and the very embodiment of our customs and culture. They have retained our heritage from the ancestors. Ancestors founded the particular land, and they have to see the land is properly maintained and ruled, and the taboos kept, festivals observed, and to see that development is going on. (Ampene interview, March 3, 1999)

The most powerful and sacred knowledge behind these rituals and festivals is considered secret. Just as chiefs are protected from the profane world by the mediation of their spokesmen (Yankah 1995), so too are powerful objects and events kept hidden and protected by indirection. Even individual elders hold knowledge about different parts of a larger ritual, with few knowing the whole sequence of rites (Asiedu Yirenkyi, interview, March 26, 1999). The secrecy of chiefly activities is accompanied by an aura of fear; many Akuapem people not connected to royal families told me that they were too scared to go to the chief's court, telling me that they were afraid of making a mistake in etiquette. In the past, one might be executed for such an offense.

The secret nature of this knowledge is noted by authors in books that make cultural knowledge public. In a popular book documenting the various festivals of Ghana, A. A. Opoku wrote in the preface that it is difficult to give acknowledgments "in a book dealing with what is sacred and to some extent, secret in our cultural heritage" (Opoku 1970). In a review of two books documenting different Akan festivals, I. E. Boama wrote:

> Two Twi festivals which every Akan should try to watch are Adae and Odwira. But there are many people who even if they have seen these festivals, they have seen only a part. For only insiders have permission to see the true activities. . . .
> If you are a child of a traditional state, buy these books to read, and once you know your nation's secrets, you won't avoid these festivals out of fear. (Boama 1954, p. 12)

Cultural knowledge, at its deepest or most true, is thus considered hidden and not accessible to outsiders; books documenting them violated that secrecy by describing rituals to non-royals and youth. The secrecy of certain historical and cultural knowledge allows powerful elders to manipulate important decisions regarding property rights and political positions, which are entwined with family genealogy and local history. As William Murphy (1980) points out about secret knowledge in Liberia, the content of the hidden knowledge does not matter as much as the privileged society (in this case, of elders) the secrecy creates.

Note that the image created here about "culture" is different from that imagined in the poem at the beginning. In the popular imaginary of culture, we see secrecy, fear, and authority. However, the poem that opened this essay, recited as part of a school cultural competition, spoke of education by elders as open and accessible to young people. Thus, the model of schooling, in which knowledge is taught to all young people of the nation, becomes the model for all learning, including that of *ammamrε*. Conceptually, *ammamrε* and *mpanyinsεm* become aligned with school knowledge; they are envisioned as being transmitted from elder to young person in an open and accessible way. In part this is due to the government's way of imagining culture, in which culture becomes the property of the nation and thus accessible to every student, an understanding of culture that I detail next.

THE PRODUCTION OF NATIONAL CULTURE

Since Ghana's independence in 1957, but more concertedly by the PNDC government that came to power in 1981, culture has been nationalized and secularized by the government. In the 1960s, under the aegis of first President Kwame Nkrumah and his promotion of African culture, cultural traditions were put on display in national celebrations such as Independence Day and other occasions, and schoolchildren and artists competed in cultural competitions in performance traditions of drama, dance, and drum language. The PNDC's effort is the first to make cultural traditions an actual subject in the curriculum.

The PNDC's cultural policy defines culture as "the way of life of a people" such as Ghanaians, stressing the mundane rather than the celebrative. The government assumes that culture promotes social cohesion and order, a functional definition of culture that makes it amenable to political use (Coe 1998). The "Cultural Policy of Ghana" (1991) states:

> Our culture is the totality of the way of life evolved by our people through experience and reflection in their attempts to fashion a harmonious co-existence between them and their environment: material and non-material. This, as a continuing process, gives order and meaning to social, political, economic, aesthetic and religious norms and modes of organisation which distinguishes us from other people. (National Commission on Culture, p. 2)

Former Secretary of Education and Culture, Mohammed Ben Abdullah, said in a published interview, "A culture is made up of the way people dress, the way they eat, the way they live, the music, the dance, their literature, their history" (Ben Abdallah 1987, pp. 15–16). These two statements form a very different notion of culture than local understandings of cultural traditions detailed above. Culture in this sense is collective, not of ethnic groups, but is a national product set against other nations (America, Nigeria) and British colonial influences. It is ordinary and everyday and thus seems available to everyone. It is also secular, in stressing social order over relationships with ancestors and spirits. This derives from an anthropological notion of culture that encompasses all social life. Such notions of culture as the product and identifying-mark of the nation have their roots in European nineteenth-century romantic-nationalism (Dominguez 1992). Through these definitions, culture becomes the product of the nation, which the government can legitimately disseminate and promote. It suggests that cultural transmission is not dependent on elders and chiefs, but rather relies on the state and state institutions, like schools. Thus, the state's policy creates new cultural experts besides chiefs and elders, namely those working for the state in cultural areas, such as bureaucrats in cultural departments, curricula writers, and teachers.

As John Meyer and his colleagues have argued, mass education is a distinctive product of the Western, modernist project, organized around a conception of the nation-state as moving toward greater progress (Meyers et al. 1992). With the universalization of the concept of the nation, so too has the idea of mass education spread, with "its highly institutionalized structure, its explicit incorporation of all members of society, its dramatic stress on individual action, and its homogeneous and universalized rationalistic frame" (Boli et al. 1985, p. 149). State-sponsored schooling is seen as vital to the development and future of the nation and as a primary way to reach the nation's future citizens, an ideal of the governmental dream (Foucault 1979; Thomas 1994). Both because children are seen as more malleable than adults and because schools are already operating, theoretically, under state control, governments, but particularly revolutionary governments, turn to schools for making and remodeling the social order.

There is a contradiction in having schools teach cultural knowledge to young people; conceptually, the modern project of knowledge-transmission in schools involves a general, common, and basic training of all citizens in the nation, specifically of children and adolescents for their future productivity. However, this notion of the school goes against the more selective local transmission of culture, in which people learn rituals and history *after* they have gained a ritual–political position and are given access to certain knowledge only after middle age. In the next few sections, I examine the transformations that occur when cultural knowledge is taught as a form of school knowledge, both in the form of that knowledge and the location of expertise.

The Transformation of "Culture" into School Knowledge

When cultural knowledge is taught in schools, it is primarily taught in the same way as other subjects, in which students sit in their seats, teachers are the source of information, and the creation of notes is the goal of the lesson. Schooling in Akuapem is notable for its teaching of practical knowledge in ways that render it abstract and into a game of word-reproduction, a litany to be learned and not questioned, with very little relevance to everyday life. Classroom teaching—and the pattern becomes increasingly clear at higher levels of schooling—consisted of the typical question–answer–response format characteristic of schools around the world; teachers asked directed questions in which the explicit goal was to elicit student knowledge but students were really supposed to figure out the answer in the teacher's mind. The discussion would result in various lists and definitions being put on the board. Then, "notes" would be given, in which the teacher would write down sentences and paragraphs on the topic on the board, often duplicating the points of the previous discussion, and students would copy these notes into their notebooks. These notes would form the basis of exercises, questions in school tests, and the nation-wide exams. Sometimes, for homework or classwork, the teacher would write questions on the board, based on the notes, and students would write the answers in their notebooks.[2] This is a labor-intensive and mechanical process. Notebooks are often the material objects around which lessons revolve: students hurry to copy notes down from the board, they are collected to be graded by the teacher who often has stacks of notebooks on his or her table, after which they need to be distributed again and corrections made. Notes are therefore an important mechanism for turning everyday knowledge into school knowledge, and verbalizing embodied knowledge through English words, definitions, and lists.

We see this process with the teaching of cultural knowledge also. In a primary five classroom, the teacher gave a lesson on "traditional occupations," which is a topic in the cultural studies syllabus for primary schools (Ministry of Education and Culture 1988). After writing "Some Traditional Occupations" on the board, he asked students to name their father's occupation. They responded: teacher, farmer, doctor, trader, dressmaker, and driver. The teacher wrote these down on the board and then checked off doctor and teacher. He explained to the students that traditional means "in the olden days," "what our forefathers were doing," before the white people came. He then got some answers he accepted; farming, trading, fishing, and hunting. He tried again, asking: What did your grandfathers do? The students responded: painter, baker, carpenter, and mason. But he was after specific traditional occupations— potter and *kente* weaver[3]—after which he felt he had enough listed on the board and ended the lesson (Fieldnotes, September 16, 1998).

In this example, the teacher's conception of "traditional" occupations did not correspond to the actual family histories of his students. Although the teacher thought he could elicit responses about tradition and the past by asking about the children's fathers and grandfathers, their answers did not correspond to his (or the Cultural Studies syllabus's) notions of "traditional." A local understanding of culture as "tradition," located in the past and not in the experience of young people, helps in the systematization of cultural knowledge in the classroom.

Similarly, in a lesson on folktales (*Anansesɛm*) in a third-year class in a junior-secondary school (ninth year of schooling), the teacher led the students through a discussion, which resulted in several lists being written on the board, namely: the stock characters in folktales; the structural parts of folktales, such as the introduction, response, story, and songs; and the virtues or morals one learns from folktales. A lesson on the Akan festival of Odwira for a third-year class in another junior-secondary school similarly produced several lists: the festivals of different ethnic groups in Ghana; the importance (functions) of festivals; and the food used to feed the ancestors at festivals. Lists are the material form by which knowledge is systematized and categorized.

Thus, as cultural knowledge is translated into different forms, it assumes a different meaning, becoming factual knowledge to be memorized and reproduced in exams. Although one might interpret this transformation of "culture" into notes, definitions, and lists as cultural knowledge becoming subject to the logic of schooling, it is also possible to see that this transformation makes the teaching of "culture" socially palatable to low-status youth. These are articulating processes: schools help in the codification and recontextualization of cultural knowledge, a form of school knowledge that is appropriate to young people. Pierre Bourdieu and Jean-Claude Passeron argue that as people of lower status gain access to prestigious occupations, those occupations lose their prestige (Bourdieu and Passeron 1990). Here, we see an analogous, but slightly different, operation; as youth gain access to a prestigious cultural knowledge through schools, a hierarchy and differentiation within cultural knowledge appears. Akuapem people continued to consider the most "deep" historical and ritual knowledge to be held by elders, while young people gained access to a recontextualized cultural knowledge that is meaningful primarily within the youthful domain of school.

The Location of Cultural Expertise

In order for the state to become involved in the transmission of cultural knowledge, the location of cultural expertise has to shift, or at least expand, away from elders to teachers or other state bureaucrats. One of the mechanisms by which teachers could attain expertise was through a series of textbooks, *Cultural Studies for Junior Secondary Schools* (1989), which teachers could use in their classes. However, textbooks are scarce in Akuapem schools; the teacher is lucky to have a copy for herself, much less one for every pupil. When I asked a teacher of Cultural Studies in an interview how he learned all the things he had described as cultural, he referenced this textbook, and then commented, "It doesn't go into much depth in there." Another mechanism is teacher training, but Akan language teachers, who are in charge of teaching culture in the secondary schools and teacher training institutions, complain about lack of student interest. Within schools, Cultural Studies is often considered unimportant in comparison to academic subjects such as mathematics and English and may not be taught. Thus, for a variety of reasons, elders continue to be legitimated as the source of knowledge about the past.

Teachers' lack of knowledge in comparison to elders may be most clear with a counterexample, in which a teacher positioned himself *as an elder* when he taught about Akan cultural traditions. Mr. Danquah[4] was a gentle man in his sixties who was always dressed neatly in a political suit—a short-sleeved, button-down shirt with pockets that hangs over matching trousers, considered as formal as a Western suit with jacket, trousers, and tie and promoted by Nkrumah. One rhetorical strategy that he used to position himself as an elder in his Akan language classes to secondary-school students was to contrast today's practices negatively with the past, a strategy that gave him authority because of his age. In a lesson about funerals, he said in Akan, "Now, it has been corrupted. It is a party," and "Funerals are not sad anymore." He emphasized the fasting that took place in the funerals of the past, and how food was only served to strangers who had traveled, not to everyone as is currently done.

He also showed his superior knowledge by giving students Akan words that they had never or rarely heard, as elders did. In the lesson on libation, a student said that a "glass" (using the English word) was needed for libation to be performed. Mr. Danquah asked for the Akan word for "glass." A girl from Akropong said, "kɔnkɔnkɔ," which Mr. Danquah corrected to "kɔnkɔ." The students responded with surprise: "Eh!" and "Ehsh!" He also gave another name for Asaase Yaa, a spirit of the earth mentioned in libation prayers, that the students had never heard before. A boy asked for the name's meaning, and Mr. Danquah explained. Thus, Mr. Danquah maintained authority through rhetorical strategies that positioned him as an elder in relation to the students; he had superior access to the more authentic ways and the purer language of the past. In fact, many colleagues and students showed him respect by calling him "ɔpanyin" or elder. Thus, in order for teachers to be seen as new experts (or as elders), they have to have more knowledge than that provided in the curriculum or textbooks. Because Mr. Danquah had a great deal of background knowledge of the Akan language and personal experience of the past, he was able to bridge the divide between school knowledge and elders' knowledge, by taking on the role of an elder.

However, unlike Mr. Danquah, many teachers do not feel comfortable or competent in teaching about culture. Some are devout Christians and avoid going to festivals, considering them dangerous to their Christian identities; therefore, they did not attend cultural activities, which is the primary way of learning about them.[5] Other teachers are strangers to the local community, and many teachers, especially at the primary and junior-secondary school levels, are young, and therefore their incorporation into ritual and political positions will happen at a later period in their lives. Thus, because of their own lack of knowledge, teachers tend to rely on student knowledge of culture in their teaching, and this resulted in tensions differing from that which normally happened in school classrooms.

The degree of teacher participation and authority during Cultural Studies, especially during enactments within lessons, depended on the teacher's knowledge of culture and the students' age. Sometimes the students, especially older ones, drew on their personal observations of funerals and festivals to reenact them in the classroom; at other times, students were dependent on direct prompting and direction from teachers standing on the sidelines or the teacher herself performed. At one end of the extreme, in one junior secondary school, the Akan-language teacher was neither from the local area nor was Akan her first language, was uncertain about her knowledge, and relied heavily on a book entitled *Akanfo Amammɛ* (Akan Culture) that she carried. Rather than teaching a lesson focused on notes and word-reproduction, this teacher asked students to enact a funeral and thus created an opportunity for them to show off their own experiential and performative knowledge. The students, especially the boys, took over the enactment of the funeral with great gusto and enthusiasm, as the teacher sat in the back of the room and glanced through her book.

> A boy who was named the organizer of the funeral (*Ayipasohene*) gave a confident little speech, standing at the front of the room. Then another boy went forward to pour libation. There were laughter and giggles as he went to the front of the room. He adjusted the cloth tied over his school uniform so that it was around his upper chest, rather than over his shoulder, a sign of respect for the spirits about to be addressed. Another boy went forward to be speech-mediator or *ɔkyeame* for the libation-pouring, but the teacher told him, *"ɛnyɛ hwee"* (don't bother) and he went back to stand at the back of the room. The boy pouring libation did not have the usual props of a calabash or bottle but he prayed with full confidence, using his cupped hands. Because there was no *ɔkyeame* responding to the prayer, other students seated in their chairs responded (*"Wɛ! Wɛ!"*) at the end of each phrase. The students laughed at themselves. Then boys beat on their desks to represent the drumming, and the girls pretended to weep in their seats. Then they rose, swept up a table representing the corpse, and they went through the corridor of the school wailing and making a commotion, bringing other students out of their classrooms. (Fieldnotes, November 9, 1999)

Although the boy heeded the teacher's comment that no *ɔkyeame* was needed, the students felt the lack of response that an *ɔkyeame* normally gives during the libation-pouring, and they took on that role, moving from observers to participants. Yet, they were embarrassed, giggling, and laughing.

Students, especially those living in royal households, were aware of their teachers' lack of expertise in these areas. As the student drummer from Koforidua representing the Eastern Region in the national competition and raised in an *ɔkyeame*'s household, commented to me: If at school he made a mistake, the teacher could tell it was a mistake but could not show him how to do it well. So, if you're not "perfect," he said, then you have to do it yourself. But at the chief's court, the elders or older people "know how to do it." So, when you make a mistake, they will take the drumsticks and show you the right way (conversation, May 3, 1999). One drummer from the royal family in Akropong who was also a trained teacher described his learning to drum in similar ways to the student from Koforidua: he had "picked it up" with his father correcting his mistakes and tapping on the table to show him how to do it right. However, he also criticized this method of informal learning: "Nobody taught us; nobody has been teaching us." In teaching drumming to students in the schools, he said that he followed "a system" and had a "developed method of teaching" (Okyen interview, September 21, 1998). So, he perceived his pedagogical techniques, possessed because he was teacher, as superior to those of the elders, and he was teaching his drumming troupe in the chiefly palace how to teach one another better. He also commented that as a teacher in a school he had greater authority than he has with his colleagues in the troupe; they did not always listen to him.

Teachers, students, and elders are all operating in a context where their expertise is constantly being evaluated by others. Mistakes in performance carry a great deal of weight in learning contexts in Akuapem. Within these contexts, positions may switch, in which students become teachers and teachers become learners, a very different scenario than what normally occurs in classrooms. Learning takes place through demonstration and imitation in ways that are similar to learning contexts outside school. Yet, at the same time, teachers who participate in chiefly activities may bring the teaching methods they use in school to transform more informal methods of learning "culture" through demonstration and imitation. Within this context of evaluating and exhibiting expertise, actors are able to highlight different aspects of their identity: as teacher, member of royal household, and elder.

THE NATIONALIZATION OF CULTURE

Since the 1960s, the government of Ghana has been engaged in organizing school cultural competitions, although they have become more formalized and institutionalized since the 1986 educational reforms. Cultural competitions began at the primary and junior secondary school level and were later extended to secondary schools. Students compete in various categories, such as drum language, dance-drama, poetry recital, choral music (usually nationalist music in Ghanaian languages), and vocational skills. If a school team wins at the circuit level, the students continue to the district level, and with continued success, go on to regional level and finally to the national competition. Teachers and students take these competitions seriously and students remember them as the main time that they were taught about "culture."

The school cultural competition attempts to appropriate local traditions and cultural practices for national ends. This is, in effect, a new tradition (Hobsbawm and Ranger 1983), very unlike festivals as practiced in Akuapem, because it is solely for children or youth and is a competition. Most town festivals have a religious significance. In the school competitions, while they begin and end with Christian prayer, the main emotional weight seems centered on the notion of the development of the nation.

The PNDC government continued this postcolonial tradition, with the innovation of introducing themes for each year's festival. In 1999, the theme was "Culture: Gateway to the Nation's Prosperity." Although at the district and regional festivals in 1999 the theme was not considered important, by the time the competitors reached the national level, the judges began using the theme to evaluate performances. By 2002, the theme, "Protecting the Environment through Culture," was emphasized even at the circuit festivals by both the district cultural officer and the judges. The theme became an important mechanism for nationalizing performing arts that are associated with local cultural traditions.

Drum language is the performance category with the most similarities to vernacular performances that occur outside of schools. Because Twi is a tonal language, a set of drums, one with a low tone and one with a high tone, can be used to "talk" relatively formulaic phrases, which can be praises or directions to the chiefs arriving that they should walk carefully. Students in the competition, drum some phrases after reciting them aloud and then the judges also test them with some phrases of their own. One difference between drumming in the chief's palace and drumming in the school competition is that during the competition, the drummer or an *ɔkyeame* (spokesperson) has to speak the phrases before they are drummed, which is not done in the chief's palace, in part because performers in the school competitions create new nationalist statements, which would be difficult to understand simply by the tonal cadences of the drumming. Another difference is that the drum text has to be written down and submitted to the judges before performance, so the drumming is not a spontaneous composition of traditional, set phrases, responding to the actual context, as is done in the chief's palace or during funerals. A final difference is that no drinks are presented to the drummers and no rituals are performed before they begin drumming. Despite the differences in context, students drew on skills acquired in community settings to perform in the drum language competition.

At the national competition, two of the best drummers had been raised in royal households and learned how to drum from their experiences at and resources in their homes and the chiefly palace (Fieldnotes, May 4, 1999). One candidate was a girl from the Brong-Ahafo region, whose drum text described the importance of culture for the nation's development and ended with these lines:

Ɔman Ghana atoto n'amammrɛ ase ama adi ayɛ yen.	The nation of Ghana has denigrated its culture for a tragedy to happen to us.
Monhwɛ amane a ato yɛn.	Look at the tragedy that has befallen us.
Ebi ne mpɛrewa nyinsɛn,	Some are the pregnancies of teenagers acting like mature women,
Akwantimfi awareɛ	Marriage to the stranger,
Nnubɔnenom ne awudie.	Drug use, and murder.
Momma yɛnsan nkɔfa	Let us go back for it,
na wosan kɔfa a, yenkyi.	And if we go back for it, we won't avoid it.
Akyea na emmui.	It is bent but not broken.
Nti yɛnsan nkɔfa yɛn amammrɛ mpasua so na ɔman no ntumi mfa ne ntoma pa mfura.	So let us go back to our culture's path, so that the nation can wear a beautiful cloth.

The other was a boy from the Eastern Region, quoted above about learning to drum, whose text consisted of appellations and praises to the paramount chief of New Juaben, Dasabrɛ Nana Oti Boateng, whose Akwantukɛse festival had been celebrated the weekend before the national school festival. The drum text began with these lines:

Dasebrɛ Nana Oti Boateng	Dasebrɛ Nana Oti Boateng,
Dwaben	Chief of Juaben
Adu Ampofo Antwi,	Adu Ampofo Antwi,
Konkorihene	Konkori
Akete-ɔnam-brɛmpɔn asum agum popa w'anim ma yɛnkɔ.	The small man who walks majestically, who has worked mightily, wipe your forehead and let's go.
Ɔkyekye akuro brɛmpɔn a ɔde ne man nam.	Great founder-of-towns who keeps his kingdom ready for defense.
Ayɔko Sakyi Ampoma Nana Yokoni,	Ayɔko Sakyi Ampoma Nana Yokoni.
Ɔsansa fa ade a ɔde kyerɛ	If the hawk catches something, it shows it.
Adakwa Yiadom Brɛmpɔn	Adakwa Yiadom Brɛmpɔn
Ɔsagyefo kasabaako a otwaa asuo barima	Savior who says one thing and does it, who banishes strong men.
Akuamoa Nana fi Dwaben Dɛeboase	The grandchild of Akuamoa from Dwaben Dɛeboase.

What he drummed was therefore far more similar to what would actually be drummed in a chief's palace. Furthermore, as a boy, he would be more likely to have the opportunity to drum there in the future than she did. The judge's public comment on this, in English, was:

> For the theme, we expected the citations to be drummed to reflect the theme of the festival. Here is a clear example cited. There is a master drummer from Eastern Region, good by all standards. But when we went back to look at the theme, then we detected his text didn't any bearing on the general theme, "Culture: Gateway to the Nation's Prosperity." Organizers, take note, and create texts reflecting on the theme. (Judge's remarks, May 5, 1999)

For the judges, it was more important for the drummers to create texts that discussed the importance of culture to the nation than to praise a local traditional ruler. The cultural performances at the competition had to be geared to the level of national concerns rather than local ones. These evaluative statements by judges carry a lot of weight because teachers and students want to win in the competitions, in order to make a name for themselves and their schools.

The school cultural competition thus involved the appropriation for "the nation" of practices and activities associated with local areas. In many ways, and as illustrated by the school cultural competition, "the nation" in Ghana is built on intense local rivalries. Towns struggle for the prize and status of development, such as electrification or school buildings, within the nation just as students competed for first prize through the different cultural performances. Despite the government rhetoric and the attempt to use the local for the nation, for students and teachers, "the nation" was a way to showcase the local—primarily their school but sometimes their paramount chief.

CONCLUSION

In the revolutionary attempt to remake and rebuild the state, the PNDC continued a process begun with independence of the state appropriating local vernacular knowledge in Ghana. In this case, the state competes against the legitimacy of chiefs and elders as the source of traditional knowledge, in order to empower the state at the expense of local authorities, a process that results in their mutual entanglement. The school curriculum attempts to make local cultural knowledge legible, systematized, socially functional, and abstract (Scott 1998), but in the process of recontextualization, renders cultural knowledge taught in school meaningful only within this youthful domain. As a result, elders, despite being simultaneously subjected to the overtures and pressures of the state, are upheld by local people as keepers of the ancestors' customs.

Although revolutionary states turn to schools because of their perceived power in influencing young people of the nation, schools as institutions have their own logic, generated historically and through the expectations and habits of families, students, and teachers. So, rather than remaking schools as Ghanaian institutions, Cultural Studies became another subject in the school curriculum, taught in the same ways as other subjects, a logocentric process for the purpose of passing all-important exams. However, as in the example of the teachers who are not considered to have cultural expertise, we also see the weight of vernacular institutions, in which the state has difficulty resignifying cultural practices as national property able to be transmitted appropriately within state institutions. In the particular transformation of "culture" into school knowledge, in fact the combination of a vernacular logic in which elders are keepers of cultural knowledge and the logic of schooling both serve to render culture as an abstract knowledge about the past; in this way, students gain access to "culture" without threatening their elders. "Culture" is thus desacralized through schooling and made appropriate for young people.

Educational reforms and political revolutions are projects; they may not succeed, but at the same time, this does not mean that nothing changes. They do open up new spaces and possibilities, both at the level of practice and that of rhetorics. The teaching of culture in schools does, at times, however briefly, open a different kind of learning space in schools, in which students learn from one another through demonstration and imitation and the teacher is also engaged as a learner, in which girls get to drum, in which young people—both teachers and students—get to exhibit their cultural knowledge and praise their town's chiefs, and in which members of nonroyal families can learn skills associated with cultural knowledge without going to the chiefly palace, a place of fear and secrecy. The state's engagement in the appropriation of local practices through schools creates a different terrain on which teachers, students, and elders can act to appropriate cultural knowledge and expertise for themselves.

This study has implications for other kinds of pedagogical practices that seek to change the social order by making students' cultural experiences visible and acceptable within schools. My work suggests that the meanings and forms of cultural knowledge shift as they are taught in a new context, such as schools. It can change students' relationships to knowledge, rendering experiences that are fluid, contested, and ongoing as more stable, abstract, and codified. Thus, we need to consider how schools incorporate students' everyday knowledge and how the move transforms the construction of that knowledge, its meanings, the location of expertise, and students' relationship to knowledge production.

NOTES

This research was funded by a Fulbright (IIE) grant (1998–99), as well as by an exploratory travel grant from the Ford Foundation's Workshop on the Problematics of Identities and States at the University of Pennsylvania (Summer 1997). I am grateful to all those in Akuapem who shared their perspectives on the teaching of culture in schools. Afari Amoako helped with translation. My thanks go to Kathleen Hall, Ilana Gershon, and Thomas Ewing who read an initial draft of this paper and made many suggestions. Any mistakes, however, are my own.

1. I am grateful to professor Mawɛre-Opoku, professor of dance and member of the Asanteroyal family, at the University of Ghana for this insight. "To *sua* [learn] is to watch something, listen to something and pick it up, try to imitate the sound, imitate what you see, imitate the habit," he told me passionately (taped interview, October 29, 1998). Mawɛre-Opoku's uncle told him, when he was a child, that if he wanted to learn wisdom, he should take a cushion to accompany the elders to their meetings (interview, December 8, 1998).

2. This teaching strategy is to some extent a response to the lack of textbooks; when textbooks are pulled out of their closets, five or more students share one book, huddled around a table, reading upside-down, sideways, or over another's shoulder.

3. *Kente* is a brightly-colored and expensive cloth made of sewed strips of woven cotton or silk.

4. All names of schools, teachers, and students are pseudonyms.

5. I discuss Christian objections to the teaching of cultural traditions in schools elsewhere (Coe 2000).

REFERENCES

Interviews

Mr. Kwame Ampene, retired teacher and founder of Guan Congress, taped interviews, in his home in Anyinasu, Central Region, March 3, 1999.

Teacher Okyen (Obuobi Atiemo Akuffo), teacher and palace drummer, taped interview, in Ofori-Boahene's house, Akropong, September 21, 1998.

A. Mawɛre-Opoku, professor of dance, taped interviews, in his office in the Institute of African Studies, University of Ghana at Legon, October 29, 1998; December 8, 1998.

Asiedu Yirenkyi, professor of theater, interview, in his office in the School of Performing Arts, University of Ghana at Legon, March 26, 1999.

Publications

Ben Abdallah, Mohammed. 1987. "Interview with Mohammed Ben Abdallah, Secretary of Education and Culture, Ghana." *Africa Report* vol. 32, no. 4, pp. 14–18.

Boama, I. E. 1954. "Reviews of *Adɛɛ* by J. H. K. Nketia and *Odwiratwa* by E. A. Tabi." *Kristofo Sɛnkekafo* 44 (September) p. 12.

Boli, John and Francisco O. Ramirez. 1986. "World Culture and the Institutional Development of Mass Education." In *Handbook of Theory and Research for the Sociology of Education*. John G. Richardson, ed. Westport: Greenwood Press.

Boli, John, Francisco O. Ramirez, and John W. Meyer. 1985. "Explaining the Origins and Expansion of Mass Education." *Comparative Education Review* vol. 29, no. 2, pp. 145–170.

Bourdieu, Pierre and Jean-Claude Passeron. 1990. *Reproduction in Education: Society and Culture*. Translated by Richard Nice. 2nd ed. London: Sage.

Brokensha, David. 1966. *Social Change at Larteh, Ghana*. Oxford: Clarendon Press.

Burchell, Graham. 1996. "Liberal Government and Techniques of the Self." In *Foucault and Political Reason: Liberalism, Neo-liberalism, and the Rationalities of Government*. Andrew Barry, Thomas Osborne, and Nikolas Rose, eds. Chicago: University of Chicago Press.

Clarke, John. 2002. "Turning Inside Out? Globalisation, Neo-liberalism, and Welfare States." Unpublished Paper.

Coe, Cati. 2000. "Not Just Drumming and Dancing: The Production of National Culture in Ghana's Schools." Ph.D. diss., University of Pennsylvania.

———. 1998. "Indigenous Scholarship on West African Vernacular Culture." Unpublished manuscript.

Curriculum Research and Development Division, Ghana Education Service. 1989. *Cultural Studies for Junior Secondary Schools: Pupil's Book 1–3*. Legon-Accra: Adwinsa Publications.

———. 1988. "Cultural Studies Syllabus for Primary Schools." Legon-Accra: Adwinsa Publications.

Dominguez, Virginia. 1992. "Invoking Culture: The Messy Side of 'Cultural Politics.' " *South Atlantic Quarterly* vol. 91, no. 1, pp. 19–42.

Dunn, John and A. F. Robertson. 1973. *Dependence and Opportunity: Political Change in Ahafo*. Cambridge: Cambridge University Press.

Folson, Kweku G. 1993. "Ideology, Revolution, and Development: The Years of Jerry John Rawlings." In *Ghana Under PNDC Rule*. E. Gyimah-Boadi, ed. Dakar: CODESRIA.

Foster, Philip. 1965. *Education and Social Change in Ghana*. Chicago: University of Chicago Press.

Foucault, Michel. 1979. *Discipline and Punish: The Birth of the Prison*. New York: Vintage.

Gilbert, Michelle. 1997. " 'No Condition is Permanent': Ethnic Construction and the Use of History in Akuapem." *Africa* vol. 67, no. 4, pp. 501–533.

Hill, Polly. 1963. *The Migrant Cocoa-Farmers of Southern Ghana: A Study in Rural Capitalism*. Cambridge: Cambridge University Press.

Hobsbawm, Eric and Terence Ranger, eds. 1983. *The Invention of Tradition*. Cambridge: Cambridge University Press.

Hutchful, Eboe. 2002. *Ghana's Adjustment Experience: The Paradox of Reform*. Geneva: UNRISD.

Kwamena-Poh, M. A. 1980. "Vision and Achievement: A Hundred and Fifty Years of the Presbyterian Church in Ghana, 1828–1978." Unpublished manuscript.

———. 1973. *Government and Politics in the Akuapem State, 1730–1850*. Evanston: Northwestern University Press.

Meyer, John W., David H. Kamens, and Aaron Benavot. 1992. *School Knowledge for the Masses: World Models and National Primary Curricular Categories in the Twentieth Century*. Washington, DC: The Falmer Press.

Ministry of Education and Culture. 1988. "The Educational Reform Programme: Policy Guidelines on Basic Education." August 17.

Murphy, William P. 1980. "Secret Knowledge as Property and Power in Kpelle Society: Elders Versus Youth." *Africa* vol. 56, no. 2, pp. 193–207.

National Commission on Culture. 1991. "The Cultural Policy of Ghana."

Nugent, Paul. 1995. *Big Men, Small Boys, and Politics in Ghana: Power, Ideology, and the Burden of History, 1982–1994*. London: Pinter.

Ninsin, Kwame A. 1991. "The PNDC and the Problem of Legitimacy." In *Ghana: The Political Economy of Recovery*. Donald Rothchild, ed. Boulder: Lynne Riemer Publishers.

Opoku, A. A. 1970. *Festivals of Ghana*. Accra: Ghana Publishing Corporation.

Owusu, Maxwell. 1989. "Rebellion, Revolution, and Tradition: Reinterpreting Coups in Ghana." *Comparative Studies in Society and History* vol. 31, no. 2, pp. 372–397.

Rathbone, Richard. 2000. *Nkrumah and the Chiefs: The Politics of Chieftaincy in Ghana, 1951–1960*. Accra: F. Riemmer.

Scott, James C. 1998. *Seeing Like a State: How Certain Schemes to Improve the Human Condition Have Failed*. New Haven: Yale University Press.

Thomas, Nicholas. 1994. *Colonialism's Culture: Anthropology, Travel and Government*. Princeton: Princeton University Press.

Yankah, Kwesi. 1995. *Speaking for the Chief: Okyeame and the Politics of Akan Royal Oratory*. Bloomington: Indiana University Press.

———. 1985. "The Making and Breaking of Kwame Nkrumah: The Role of Oral Poetry." *Journal of African Studies* vol. 12, no. 2, pp. 86–92.

5

Folk Schools, Popular Education, and a Pedagogy of Community Action

William Westerman

What can make pedagogy revolutionary is not just the content, but the process, and the question of who is the teacher and who is the student. There are the dualities—reading/writing, listening/speaking, answering/questioning, accepting/investigating—but in much formal schooling, equal emphasis is not given to both halves of an engaged communicative process. Conceivably, though not always, teaching and studying can exist in a dialogic relationship that in itself is a revolutionary reformulation of the standard classroom technique. Beyond that, occasionally the educational process can lead to further action and to social change. This chapter concerns itself with the act of study as a proto- and prerevolutionary act, an act of questioning and an act of challenging the existing social order. More accurately, this essay addresses historical examples when the act of study was a force advancing a revolutionary process. In fact, the kind of pedagogy I discuss is one in which the student's actions and questions, and the authority of daily life, are given a primacy that they do not have in static educational models.

In particular, this chapter focuses on two strands of revolutionary pedagogy—one that originated in the folk schools of Denmark and took root in the mountains of Tennessee, influencing the Civil Rights movement as well as generations of labor organizing in the United States and one that originated in the slums of Brazil, took a detour through Portuguese West Africa, and took hold in revolutionary and religious communities in Central America and parts of South America before moving on to the wider developing world. These two strands, the northern thread spun by Danish nationalists and Myles Horton and the southern spun by Paulo Freire and Amilcar Cabral, intertwined briefly before the deaths of Horton and Freire in the 1990s. What ties the two together are themes of freedom, empowerment, and emancipation, on the one hand, and techniques of developing literacy, orality, and knowledge of the community, on the other. It turns out that an understanding and an appreciation of folk culture is central to both. In the conclusion, I attempt to braid these threads with a third strand, that of folk-life studies, to examine the qualities inherent in a discipline that studies popular culture which make it so suitable for an activist pedagogy.[1] It will be my contention that there is something about these three pedagogical disciplines that in combination make for an education that is truly revolutionary: critical literacy, orality (not in the sense of rote recitation but in the sense of mastering oral skills and being able to wield the spoken word), and knowledge, including self-consciousness, of one's folk culture.

THE DANISH FOLK SCHOOLS
AND THEIR INFLUENCE

One of the manifestations of Romantic nationalism in Europe was the collection and study of folklore materials, in such nascent nations as Finland, Italy, and Germany. Following the philosophical influence of Herder, proto-folklorists roamed the countryside, collecting songs and tales (*Märchen*), from peasants, in the process elevating and romanticizing the peasant class within their own society while asserting the nationhood of their own people. But all was not just a matter of nationalism. Another feature of this elevation concerned the education of the peasantry as part of the process of civilizing them and preparing them for political participation in the rising nation. As the influence of monarchies waxed and waned in the decades following the French Revolution, and as national boundaries were being drawn and redrawn, it was also important that the peasants were kept clear about their national and linguistic allegiances. Political and intellectual elites in smaller European countries were concerned about losing national identities and wanted to bolster claims of national autonomy. In the wake of the Napoleonic Wars, devastated nations looked to assert national cultures once more. Some middle-class intellectuals harbored democratic impulses, too, and after the revolutions of 1848 looked to the peasant class as future citizens in post-feudal constitutional societies with bicameral parliaments, at least one chamber of which was to be popularly elected.

In the Scandinavian countries and in Denmark in particular, the task of educating the peasant was part of the work of establishing cultural and linguistic identity. In 1830, Henrik Wegeland, a Norwegian poet, wrote a theoretical essay called "Encouragement to Country Folks to See to Their Own Education" (Paulston 1974, p. x). Similar ideas were developing in Denmark, which had lost land to Germany and which was faced with the spreading influence and increasing use of the German language by its southern elites (Begtrup et al. 1926, pp. 94–95; Coe 2000; Rørdam 1980, p. 40). A Lutheran minister who also happened to be a student of Norse mythology, N. F. S. Grundtvig (1783–1872), developed the idea in 1834 that one remedy was a form of popular education not just for elites but for the peasants as well, those who "have to feed themselves and the officials too" (Adams 1975, p. 20; Manniche 1939, p. 84). In that way, while not challenging the idea of monarchism with democracy, at least the rural folk could participate in the strengthening of a Danish nation. Grundtvig established the first of these

"Folk Schools" ten years later, in 1844 (Manniche 1939, p. 87; Paulston 1974, p. x). His school failed, but a second, more successful school was founded by Christian Kold (1816–70) seven years later with one-tenth the initial start-up money (Begtrup 1926, pp. 95–99). Together these schools spawned a movement that over the course of the next 80 years educated some 300,000 Danish citizens (Adams 1975, p. 20). By the mid-1920s, as many as 30 percent of the rural population of Denmark had attended a folk school (Hart 1926, p. 41).

Directed at farmers, peasants, and rural artisans, the Folk High Schools ultimately offered a curriculum consisting of Danish language, history, and social conditions, Norse mythology and literature, agricultural studies, music, and physical education, in combination with (literal) "field work," time spent developing better manual agricultural and domestic skills. More hours were spent studying history and social conditions than any other subject. There were no books; everything was oral, including poetry and song (Adams 1975, p. 20; Hart 1926). Grundtvig believed that "Academic life tends peculiarly to lead bookish men into false paths unless it is continuously corrected by an education that comes not out of books, but out of the life and the work of the people" (Hart 1926, p. 106). His work was furthered by his disciples Kold and Ludvig Schrøder, who "believed in 'the poetry of human activity, the romance of daily work' " (Manniche 1939, p. 95). All believed strongly that education had to have application in everyday life. Wrote Grundtvig, "Dead are letters even if they be written with the fingers of angels, and dead is all knowledge which does not find response in the life of the reader" (Manniche 1939, p. 83).

Grundtvig's belief was shared by many Romantics in nineteenth-century Europe. Leo Tolstoy wrote in 1862 "the home conditions . . . the field labor, the village games, and so forth . . . are the chief foundation of all education . . . Every instruction ought to be only an answer to the questions put by life" (Tolstoy 1862, pp. 14–15). But folk schools maintained an explicitly activist agenda as well: "Facts enlarge life when they are *in* life. And the task of education, as of civilization, is that of finding some way of getting the facts we know used in the reordering of the world," wrote one American disciple (Hart 1926, p. 85).

Grundtvig's overarching operational theory in his "school for life" involved three factors involved in this definition of folk school education as an approach that, in the words of one modern interpreter:

> (a) arises from life, i.e. it is rooted in the life of the individual and the people. It speaks into the life situation of the young and it is inspired by the people's myths and history (b) is living. That is to say . . . not based on the book, but on the living, spoken word, above all conversation . . . (c) aims at life. The education has the task of shedding light both on human life as such and on the individual's life. (Bugge 2001, pp. 65–66)

Above all, unlike much state-run education, folk schools were "not examination schools. Their aim is not to enable the student to pass a particular exam" (Bugge 2001, p. 67).

Language, culture, and music held special roles in Grundtvig's thought and, later on, in the schools. Grundtvig had traveled to England three times between 1829 and 1831, to study Anglo-Saxon texts, including Beowulf, which he copied to bring back to Denmark (Begtrup et al. 1926, pp. 84–85). He also translated works from Latin into Danish, because of the importance of "a living voice speaking the mother tongue" (Begtrup et al. 1926, p. 83). As for music, a later American visitor wrote, "It is a delightful experience to hear the students sing hymns, ballads, folk songs, and patriotic songs. All of them sing and most of them sing well" (Knight 1927, p. 81). Eventually, a high school song book became part of the folk school curriculum, and went into at least 16 editions by 1974; the largest included over 800 songs (Rørdam 1980, p. 110).

The folk school movement lay behind many political and economic changes in Denmark during the late nineteenth century, including the development of agricultural cooperatives, land reform, and the rise of participatory democracy through Left parties and politics in general, including successively larger electoral victories by Leftist parties in 1872, 1884, and 1901 (Knight 1927; Rørdam 1980). Not all the schools were politically radical (Whisnant 1983, pp. 130–136), but they had an impact on the economic structure of Denmark, at least. One American visitor, education professor Edgar Wallace Knight of the University of North Carolina wrote admiringly that:

> Since 1864 [Denmark] has restored herself, has developed an effective system of general education and of adult and agricultural education, perfected an agricultural system that is unexcelled anywhere, created coöperative agencies which are the marvel of Europe and America, removed illiteracy among the people, converted tenants into owners of homes, organized the farmers and enabled them to protect their interests, and brought to the masses of the Danish people a high level of intelligence and culture and material prosperity. Education has transformed rural Denmark. The right kind of education properly directed can transform any rural community in the United States. (Knight 1927, p. 10)[2]

The Danish Folk Schools showed remarkable longevity, and were still expanding into the twentieth century. In 1921, an International People's College opened in Elsinore, attracting the attention of people far beyond Scandinavia (Manniche 1939, p. 21). American writer Joseph Hart, in a glowing profile of the folk schools published in New York entitled *Light from the North*, captured the essence of the folk school philosophy in this description:

> [The Danes] do not create the minds that develop in those schools. They do not control those minds. They do not tie old forms of culture to those minds in order to make sure that they will be cultured later on. They provide the proper soils for growth; they take young people at just the right season of their lives; they cultivate them intelligently; they surround them with the climates of cultural growth; they trust the processes of development; they provide that "silence and long time" which John Keats held to be foster-mother of culture; and, for the most part, they reap the expected harvest in due time. They are culturers of minds; they work with nature; and they get the rewards that intelligence deserves. (Hart 1926, pp. 78–79)

The education, as Hart saw it, was in the cultivation process, not, metaphorically speaking, in the crop or the seeds themselves. Later on, there would be specific folk high schools devoted to Danish handicrafts (founded in Kerteminde in 1952), language (founded in Kalø in 1952), and art (founded in Holbæk in 1965) (Rørdam 1980, pp. 183–185).

Several American educators in particular were impressed enough by the idea to travel to Denmark and use the folk schools as a model in their own work. Olive Dame Campbell authored a 1928 book on the topic (Campbell 1928) and then went on to found the John C. Campbell Folk School in the Appalachians (named after her husband), with which folklorists might be familiar, since it has been discussed more critically by David Whisnant in his book *All that is Native and Fine* (1983). In her study, Olive Dame Campbell had noted that Grundtvig "emphasized the importance of an experience in the common life and labor of the everyday man" (1928, p. 58) and she included the study of social and economic conditions in her curriculum, though she avoided some of the more politically radical elements of the Danish schools' analysis (Whisnant 1983, p. 136).[3]

THE HIGHLANDER SCHOOL

The influence of Danish Folk Schools in the United States was spread even more widely through the efforts of Myles Horton (1905–90), who spent a year in Denmark in 1931, following the onset of the worldwide Great Depression. Born in mountainous Savannah, Tennessee in 1905, Horton came from a poor rural background. Though his grandfather had been illiterate, his parents stressed education, and after college, he had ended up studying at Union Theological Seminary under Reinhold Niebuhr, "to try to find out," in his words, "how to get love and social justice together" (Horton 1990, pp. 32–35). He followed this with a year of studying sociology at the University of Chicago, and while in Chicago Horton met Jane Addams and participated in programs at Hull House (Horton 1990, pp. 47–48). He first became introduced to the folk schools after attending a community folk dance session at Reverend Aage Møller's Danish Lutheran church in Chicago. Møller and another pastor, Enok Mortensen, told Horton about the Folk School movement and suggested he visit Denmark (Glen 1996, p. 16; Horton 1990, p. 50).

Horton would also likely have been aware not only of John C. and Olive Dame Campbell's work with folk schools, but with more political interpretations being published in the United States as well. Knight's study, cited above, concluded with this declaration:

> Too many of the farm workers in the United States are in the toils of the vicious system of tenancy, owning not an inch of the soil they live on nor a single shingle in the roofs under which they sleep. The Danes have been able through education and coöperation to reduce social injustice and to increase the well-being of all the people. (Knight 1927, p. 236)

Such sentiments would have been very attractive for Horton, who was trying to figure out at that time how to put his ideals into practice. Indeed, at the Danish Folk Schools, Horton was drawn by the sense of community, by the belief that school was a social experience, and by the appeal of placing home life in its larger context. Students, he wrote, "could learn not only what to do at home but have a chance to study social forces at work" (Horton 1989, p. 30). The radical potential of "these democratic schools" and their Danish founders also impressed Horton, who later wrote admiringly of Grundtvig's "creative powers . . . and . . . emotional energy . . . which enabled him to strike out, almost single handed, against the economic and spiritual poverty that enslaved the people" (Horton 1944, pp. 23–24).[4]

He returned to the United States in 1932, interested in education in a local community setting, like the folk schools, rather than mass educational reform where results were harder to gauge. Being tied to a community was paramount (Horton 1938, pp. 280, 295; 1990, p. 56). Enlisting Niebuhr's help in the fundraising and organizational effort, he planned to open "The Southern Mountains School" in North Carolina, "for the training of labor leaders in the southern industrial areas" (Horton 1990, p. 61). Thus, unlike the Danish Folk Schools that had not been overtly concerned with the political process, Horton introduced into his concept of folk school the express purpose of effecting political change in the region. This signaled a switch in the development of folk schools, in the words of one historian, after which they came "to secure generally liberating rather than adjustive outcomes" (Paulston 1974, p. ix). But Horton's view was also holistic; he wrote shortly after founding Highlander that "Schools must combine the economic, social, intellectual, esthetic, and moral elements of our culture, just as ordinary people combine them in ordinary life" (Horton 1938, p. 267).

Horton intended his school to be the first of many such schools in the region with this social agenda. "The Southern Mountains School"—a name initially chosen strictly for fundraising purposes—Horton and his associates renamed the Highlander, choosing a term, interestingly enough, that had been coined and made popular by John C. Campbell (Campbell 1921; Horton 1990, p. 63). The Monteagle, Tennessee location was selected for the very practical reason that Dr. Lillian Johnson, a former college president and student of John Dewey, donated her land to the project (although threatening to withdraw support because at first she found the enterprise too radical) (Horton 1990, pp. 63–65). The first two classes offered there were in psychology and cultural geography, a discipline close to, if not overlapping, folklife studies. Niebuhr, Joseph Hart, and Norman Thomas were among the school's charter advisory board members (Horton 1989, p. 46).

Over the next three decades in particular, Highlander sharpened its mission. The first students and trainees concentrated on labor organizing, including building unions and planning rallies. Highlander even became an official training center for the Congress of Industrial Organizations (CIO) in the South, for example (Horton 1990, p. 87). Yet, it was in the education and training of activists for integration wherein Highlander came to national prominence, as the cradle of the Civil Rights movement. Rosa Parks, John Lewis, and Martin Luther King, Jr. were among the most prominent of hundreds of Civil Rights activists who received training at Highlander, one of the few integrated institutions in the entire South.

Three aspects of the curricular philosophy at Highlander were particularly important and novel within the American context. The first was the idea of experiential education that Horton clearly adapted from Grundtvig (and Tolstoy and Jean Jacques Rousseau). People learn by experience in real-life settings:

> If we are to think seriously about liberating people to cope with their own lives, we must refuse to limit the educational process to what can go on only in schools. The bars must come down; the doors must fly open; nonacademic life—*real* life—must be encompassed by education. Multiple approaches must be invented, each one considered educative in its own right. (Horton 1973, p. 331)

However, Horton expanded the concept of experiential education from the Danes' emphasis on agricultural fieldwork and artisanship to include the experience of democracy, cooperative economics, and participation in labor strikes within the learning environment (Horton 1990, pp. 68–69). The Highlander Folk Cooperative, a cannery, opened in 1934, while participation in strikes, including having the students join picket lines at major actions in the region, was also one of the earliest features of the school (Horton 1989, p. 54).

Horton also went further than the Danish Folk Schools in the belief that the people, including students, held the answers to their own problems based on the experiences of their lives. Horton later described this change in his point of view:

> [W]e saw problems that we thought we had the answers to, rather than seeing the problems and the answers that the people had themselves. That was our basic mistake. Once you understand that, you don't have to have answers, and you can open up new ways of doing things. (Horton 1990, p. 68)

The method was in the ability, as he wrote, "to learn how people learn, and respect what they already know" (Horton 1990, p. 69).

Horton's cultivation of leadership was also one of Highlander's greatest strengths and innovations. He did not articulate this much in his published work, but occasionally revealed the philosophy behind his method:

> It is, of course, impossible to bring together more than a small percentage of the people, but this is all that is necessary. Leaders are aroused people, who, like the aroused atom, start a chain reaction which is self-propagating. It is the multiplication of leadership that gives power to the people and enables them to make reality of dreams . . .

And then, with a nod to folklore:[5]

> The key that will release the people's power will not be found in the hands of a Paul Bunyan or a John Henry, but in the hands of the Johnny Appleseeds. (Horton 1947, p. 82)

The incorporation of cultural activities into the curriculum, the third significant and novel element of Highlander pedagogy, was part and parcel of the
educational as well as political process. Horton got the idea of community
singing from Grundtvig and the Danes and dancing from his time in Chicago
(Horton 1973, p. 329; 1983, p. 29). In personal notes written around 1931,
he suggested "School and community life should be built around get-togethers
such as singing, dancing, discussions, dinners, etc." (Horton 1989, p. 30). As
he noted on the use of crafts in elementary education, "Handicraft and art-
work should be presented in such a way that beautiful things would not be
thought of as unattainable luxury, but as necessary parts of life," and that
knowledge "of local trees, flowers, herbs, and wild life" was an important
part of this curriculum, too (Horton 1938, p. 287).

Horton's first wife, Zilphia Johnson Horton (1910–56) has only recently
been recognized as a major figure in the pedagogy of Highlander, and has
not yet been fully recognized for her significance in American cultural work
and folklore (Carter 1994; Glen 1996; Horton 1989). She came to the
school in 1935 and introduced singing into the school curriculum (Horton
1990, p. 75) as well as the off-campus work of labor organizing and striking
(Carter 1994, p. 8). Myles credited her with adding a new dimension to the
school: "I learned a tremendous lot from Zilphia, my wife, who brought in a
whole new cultural background, drama and dance and music, oral history,
storytelling—all kinds of things that I'd grown up knowing but just hadn't
thought of as being related to learning" (Horton and Freire 1990, p. 41).
She collected folk and labor songs, transformed and wrote new lyrics for others, composed new ones, and encouraged the students to do the same (Glen
1996, pp. 46–47, 66; Horton 1989, p. 120). One book of her collected
songs had an introduction by John L. Lewis, president of the United Mine
Workers (Adams 1975, pp. 72, 225). Students would go out to picket lines
and collect songs from one strike, and often, later on, transmit those songs to
strikers on other picket lines, thus acting as agents of transmission in the labor
folk tradition. Zilphia Horton later said in an interview, "The people can be
made aware that many of the songs about their everyday lives—songs about
their work, hopes, their joys and sorrows—are songs of merit" (Adams 1975,
p. 76). Singing was also popular on the school grounds, and the BBC even
broadcast a concert from Highlander in 1937 (Glen 1996, p. 47). It was
at Highlander that "We Shall Overcome" was adapted by Zilphia Horton in
1946 to become the anthem of an embryonic political movement; she adapted
the words, slowed down the tempo, and provided accordion accompaniment
(Carter 1994, p. 15). Other songs of Civil Rights—"We Shall Not Be Moved,"
"This Little Light of Mine," and "Keep Your Eyes on the Prize"—were first
associated with the movement there during her time, whether in their original
or rewritten forms (Glen 1996, pp. 176–177).[6]

The cultural component of the Highlander experience continued with the
arrival of folklorist, collector, and singer Guy Carawan in the 1950s, and later
his wife Candie Carawan (they eventually became the directors after Myles
Horton's death), and frequent visits from Pete Seeger. But participation, not

performance, was the key. The Civil Rights activist, SNCC chairman and later Congressman, John Lewis wrote, "Besides the workshops and the speakers and the discussions, we did a lot of singing at Highlander. It didn't matter whether you could carry a tune or not, everyone sang. Even me, and I *cannot* sing" (Lewis 1998, p. 90). Social folk dance was equally a part of community building, as it came to be in Denmark too, and square dances were a regular feature of life at Highlander. Labor dramas were developed or written by staff and students for tours of the South, perhaps as many as a hundred in all (Adams 1975, p. 72), and largely under Zilphia Horton's direction, as she had studied drama and advocated it as a technique for social change (Carter 1994, p. 10). The school also encouraged worker-students to set down their first person narratives of life in the labor movement, and many of these were published in the school newspaper or in mimeographed anthologies (Horton 1989, p. 121). Myles Horton would later sum up, "Ballads, hymns, folk songs, songs of protest: all these have done much to arouse people to awareness and to the sense of community" (1973, p. 330). In many of these areas, Zilphia provided the initial spark.

Both the Danish Folk Schools and the Highlander School took as their starting point the education of rural adults, educating them beyond basic skills and encompassing critical thinking and participatory democracy. If a radical message was implicit in the work of some of the Danes, Myles and Zilphia Horton made that a more explicit and important part of their work. In that sense, the work of Highlander always strove to be revolutionary, reaching its epitome in the Civil Rights movement of the 1950s and 1960s. At the same time, to the south, Latin American revolutionary educational movements were also developing, and they too would become part of larger social revolutions.

PAULO FREIRE, LITERACY, AND POPULAR EDUCATION

A contemporary of Horton, Paulo Freire (1921–97), embarked on a similar path and concerned himself with adult education and literacy throughout Latin America. He did not learn from the Danish Folk Schools directly, but participated in political movements of the poor in his native Brazil. Even more than Horton, Freire developed a specific pedagogical methodology that was to have an impact on a political plane that reached higher than protest into the very workings of revolutionary government. Freire's work began in Brazil, but as a result of political circumstances, his temporary exile, and his travels later as an acknowledged innovator, his ideas were put into practice in Chile, Central America, particularly Nicaragua, and three formerly Portuguese colonies in Africa as they reached independence, as well as in poor neighborhoods and literacy campaigns throughout the Americas. As with Horton, Freire's work began with seemingly basic skills but placed them within a context of culture and the recognition of prevalent socioeconomic conditions.

Freire's now well-known method involves a radical reformulation of the teacher–student relationship. In formal educational models, the teacher traditionally holds all the knowledge, and deposits information into students, who function as mere receptacles. Freire introduced a more critical model of the educational relationship, which recognizes the role of the student's life experience in making sense of the surrounding social reality. The student's understanding and critique of social reality not only become part of the educational dialogue between student and teacher (since all learning, according to Freire, is based in a dialogic relationship) but also becomes the concrete bases for the teaching of literacy skills. The student's life becomes part of the curriculum. The student learns to read not meaningless phrases with no social context, but phrases with a bearing on everyday life experience. Thus, the student is learning to read the world in addition to the word, to use Freire's phrase from his seminal work, *Pedagogy of the Oppressed* (Freire 1970). In other words, what sets Freire's method apart is that the student becomes an active participant in the educational process, through questions, dialogue, and the introduction of life experience; the teacher does not have all the answers, and the content of the reading primers are no longer irrelevant phonetic phrases, but sentences that reflect on the everyday life reality of the student.

Freire pioneered these methods in Brazil (and later Chile) in the late 1950s and early 1960s. Because of his literacy campaigns and outspokenness against the government, he was jailed and eventually exiled by the Brazilian dictatorship. But Brazil was hardly the only country involved in adult literacy campaigns or in the analysis of social conditions during the 1960s, 1970s, and 1980s. Because of its sheer size, it was harder to have a nationwide influence there. In fact, the first great national literacy campaign in Latin America began in Cuba on April 15, 1961, ironically the very day of the Bay of Pigs invasion (Kozol 1978, p. 354).[7] As with Freire's method, which begins with the analysis of a drawing by the students, the Cuban lessons began with the social analysis and discussion of a photograph (Kozol 1978, p. 353). In both cases, the teaching of literacy begins with the social realities of the students.

In a study of the folk school movement in Bangladesh, the Danish scholar of folk schools, K. E. Bugge discusses Freire's work as the natural outgrowth of Grundtvig's method played out in the economic context of developing countries. He finds extraordinary similarity in Grundtvig's and Freire's outlook. Both were concerned with empowerment, emancipation from oppression (be it cultural or economic), equity,[8] and most of all, freedom (Bugge 2001, pp. 69–73). Methodologically both were similar as well, emphasizing education drawing on life experience, and the importance of both literacy and oral expression. As Bugge notes, both Grundtvig and Freire believed "that backward social groups have lost the urge and ability to express themselves orally" (Bugge 2001, p. 74).

The most widespread and successful application of Freire's approach took place in Nicaragua beginning in early 1980, where a revolutionary government sought to use literacy to extend a political revolution into a social one. In fact, during a visit to Nicaragua three months after the Sandinista revolution that toppled Anastasio Somoza, Freire reportedly "stressed the importance of providing opportunities for learners to practice their creativity" (Cardenal and Miller 1981, p. 17).[9] As in Cuba, a literacy campaign was one of the first initiatives of the Sandinista revolutionary government.[10] In the 1980 "Crusade," illiteracy was lowered from approximately 50 to 13 percent nationwide and it has been estimated that "more than one-fifth of the population participated directly in the campaign" (Berryman 1984, p. 237; Chacón and Pozas 1980; Miller 1985, p. 200).

As in Cuba, the reading primer was developed around thematic chapters that reflected the lives of peasants and fishing communities, rather than readings with no social context. The first reader included such topics as a biography of Sandino, the "struggle for national liberation," "exploitation by foreign and national elites," the "rights and responsibilities of the new citizenry," land reform, health care, and women's rights (Arnove 1986, p. 22). Words introduced in the first reader included *revolución* (revolution), *liberación* (liberation), *genocidio* (genocide), and *masas populares* (popular masses) (Arnove 1986, p. 22). The official guidelines indicated that a central goal was "to actively engage students in the observation, interpretation, and analysis of their life circumstances" (Arnove 1986, p. 50), and specifically promote a basic understanding of "the national development plan, and the emerging political and economic structures" of the postrevolutionary society (Cardenal and Miller 1981, p. 6). One science textbook included the following statement on pedagogy: "We can only say we have learned something when we are capable of applying [knowledge] to transform little by little our reality" (Arnove 1986, p. 50). In other words, this pedagogical style was a three-part process, involving "discovering reality, interpreting it, and transforming it," while "the learner . . . also teaches and the teacher . . . also learns" (Suarez quoted in Miller 1985, p. 94).

Moreover, the *brigadistas* who were deployed in what was also known as "the cultural insurrection" to teach literacy, also became involved in manual labor, farming, craft production, and other efforts to integrate theater, dance, oral poetry, and festivals in the countryside into the literacy campaign (Cardenal and Miller 1981, pp. 8, 21; Chacón and Pozas 1980; Whisnant 1995, pp. 239–240).[11] Some 450 students were enlisted in "cultural brigades" that did not participate in literacy teaching but who were deployed to study culture and collect oral histories and other folk cultural knowledge (Miller 1985, p. 67). Fernando Cardenal, the Minister of Education, wrote in December of 1979 that there were plans to include as part of the literacy campaign the collection of national songs, refrains, folktales, and legends (Cardenal 1981, p. 35).

Most ambitiously, perhaps, Cardenal announced a plan to record two thousand oral histories of the insurrection to be deposited in a future "museum of oral history of the struggle for our freedom" (Cardenal 1981, p. 36). Within five months of the commencement of the campaign, some 7,000 oral history interviews had been conducted (Equipo DEI 1981, p. 144), though it is not clear whether all were recorded on audiotape. One Ministry commission later wrote, at the second congress reviewing the progress of the campaign in September 1980, that projects to develop a cultural atlas, collect information on folk medicines, and to incorporate art "as an instrument to deepen the mechanism of learning" would be included (Equipo DEI 1981, pp. 226, 229). The commission also announced plans to develop a National Museum of Literacy, with exhibits on literacy, including many essays and poems written by newly literate peasants, as well as collections of "our popular [folk] songs and our legends" (Equipo DEI 1981, p. 238). Eventually an Institute of the History of Nicaragua was founded under the Sandinistas in 1987, and all the materials from the Literacy Crusade and the oral history project, including cassette tapes, are archived there, at what is now the combined Institute of the History of Nicaragua and Central America, located at the Universidad Centroamericana in Managua.

The Nicaraguan case was clearly revolutionary, not only politically, but also in the sense that educational work and cultural work were literally brother disciplines. The literacy campaign was in many ways the centerpiece of the Sandinista goverment's initial work and arguably its most successful project, as it had been in other countries where Freire had been involved (see, e.g., Freire 1981). Not only was it the most successful application of Freire's method in terms of scope and short-term results, but also its qualitative impact went far beyond literacy and accomplished more than the reform of an educational system. While the idea of the literacy campaign was top-down (as it was with the Danes and Olive Dame Campbell), Freire's method is, to the fullest extent possible, dialogic and bottom-up, in his words, revolutionary rather than reformist (Cardenal and Miller 1981, p. 17). In Nicaragua this approach was put fully into practice during an intensive half-year period. What may be surprising is how integrally the understanding of popular culture was interwoven in that pedagogical process.

Freire's method was also applied in the teaching of adult literacy in El Salvador, during its attempted revolution in the 1980s. In contrast to Nicaragua, the Salvadoran campaign did not have the backing of a government newly in power, and education took place in villages, guerrilla camps, and "liberated zones" (Hammond 1998), carried out by the guerrillas as well as by nonviolent political activists in the so-called popular movements. Among the first syllables taught to Salvadoran peasants were "*cam-pe-si-no*" (the Spanish word for "peasant"). This was very much based precisely on the model not only from Freire but also from the war of independence in Guinea-Bissau and Cape Verde advanced by Amilcar Cabral (1924–73), whom Freire had influenced personally, and who in turn influenced Freire's later work as well.[12]

At the same time that literacy campaigns were underway in Nicaragua, El Salvador, and elsewhere, another movement, Liberation Theology, with similar aims and methods was becoming widespread throughout Latin America. The emergence and ideals of this movement showed the influence of Freire's approach beyond literacy and into the realm of religion. A radical reformulation of church teaching, favoring interpretation from the perspective of the poor and oppressed, Liberation Theology developed as an outgrowth of Vatican II by theologians, clergy, and laity concerned about social conditions in Latin America, including Brazil, Peru, and Central America. The method for disseminating this approach came in the form of so-called Christian Base Communities, discussion groups among urban and rural poor in which doctrine was discussed and debated in the context of the social conditions of people's lives, not as dogma. The structure of these Communities was similar to the structure of the literacy classes, with a high value placed on critique, questioning, life knowledge, and the reassignment of intellectual authority. Literacy, religious practice, and culture were part of a holistic social transformation, employing similar methods even though on a surface level they might appear to be distinct.

Typically, Base Community leaders would begin by reading passages from the Bible and prompting discussion, placing authority for interpretation in the hands of the parishioners, rather than in defending their own unquestionable authority.[13] Base Community work could also involve the exchange of knowledge between the community members and the leaders. It could really be said that the Base Community experience was an even less formal kind of popular education than literacy campaigns or folk schools, largely because its social structure was so radically different. According to one Salvadoran former Base Community lay worker I interviewed in 1985, "70% of the whole [popular political] movement in El Salvador" had come from a background "in a base community, religious base community." He described his experiences as a leader:

> There is this ambience, this feeling, I mean, the situation in, when you come into the area, when we were coming into the neighborhood, into the shanty-towns, like we feel that, you know, there was the place in which we will have to be. Because when we were coming in: "Hello, how are you?" and everyone [came] to hug us, you know, and to say, "How you been?" and "Listen, I have this mango here . . . Do you want the half? I'll take the other half, I haven't eaten," you know, and there is all, all the situation of sharing *everything*. Everything.
>
> Yeah, it's a community of sharing, of giving. Rather than you know receiving, just receiving. . . .
>
> We sat in different groups . . . Usually in each group there were at least ten people, and we were five, seven people going at the time. We were talking about 70 to 100 people many times, sitting you know, underneath a tree, to the shade of the tree, on roads, on the, you know, on the land, on the grass . . .
>
> It wasn't a lecture. It never was a lecture. Because in order to [make] people really realize, and to make—not just realize but to make them understand and that from this understanding could [one] take a step, it had to be a discussion, it had to be a, you know—like, I learned a lot from them, a lot. Even, you know, the way that plants have to be grown, and what type of help have to be used to cure different things. I mean that type of things. . . .
>
> It's very incredible. I mean, the first time that I went over there, you know, I was study—in my . . . senior high school and thought, I thought, you know, I'm going to, to *teach* them. And it was the opposite. I went to learn. Not to learn about the herbs, and stuff, but to learn how they were managing to live

with those conditions. From the first time that I went over there, that was it. That made it. Everyone changed from that first time . . .

[We were] between 18 years, 30, I think 30 years. Young people, I think . . . We were really young, most of us. And the people that we were talk-ing to were old, young, and children . . . Many times there were the whole fam-ily in one group. (Rolando Interview, May 29, 1985)[14]

This Base Community model illustrates, in a way, Freire's more secular notion of education as joint process:

> Every thematic investigation which deepens historical awareness is thus really educational, while all authentic education investigates thinking. The more edu-cators and the people investigate the people's thinking, and are thus jointly educated, the more they continue to investigate. Education and thematic inves-tigation, in the problem-posing concept of education, are simply different moments of the same process. (Freire 1970, p. 101)

The Base Communities, particularly in Central America, were applying this approach in religious terms, and anticipating an outgrowth into social action.

While literacy campaigns involved reading primers and even photographs as a basis for the social analysis that lead to an enriched literacy, Base Communities used Biblical texts, particularly the Gospel, as the starting point for a discussion of social inequity and injustice. While the movement certainly produced its share of written theology, it is also noteworthy for some of the well-documented examples of community life and religious interpretation, particularly Ernesto Cardenal's four volumes of base community transcripts, *The Gospel in Solentiname* (Cardenal 1982).

This approach, which became known also as the popular church ("popular" meaning "people's" but also being the closest equivalent in Spanish of the English word "folk" [see Paredes 1969]), was part of the larger scope of people's movements, including popular literacy and adult educational cam-paigns, popular organizing and labor movements, guerrilla armies, and pop-ular music, such as *Nueva Canción* (the "New Song" movement originating in Chile and spreading beyond the Andes throughout Latin America), with an overt political message. Such movements began in the countryside with peasant populations, but also included the urban poor. There was an open acknowledgment within these movements that the cultural, the political, and the pedagogical were closely connected, and that culture was not something that belonged only to an elite few.

It is important to recognize that popular culture was part of these grow-ing movements from the earliest days of Freire's work in the early 1960s, and that, as was the case at Highlander, music and art were integral to this kind of work. As Freire wrote in an autobiographical essay:

> It is no accident that the words culture and popular were so frequently present in the movement's vocabulary. Popular Culture Movement. Cultural centers, Cultural circles. Cultural squares. Popular theater. Popular education. Popular

medicine. Popular poetry. Popular music. Popular festivals. Popular mobilization. Popular organization. Popular art. Popular literature.

One of the stated objectives of the movement was the preservation of popular culture traditions: the people's festivals, their stories, their mythic figures, and their religiousness. In all this we found not only the resigned expression of the oppressed but also their possible methods of resistance . . . none of th[is] ever escaped the movement's notice. (Freire 1996, p. 117)

More specifically, he wrote upon a visit to Port-au-Prince that he saw the local popular painting as a crucial expressive form:

It was as if the Haitian popular classes, forbidden to be, forbidden to read, to write, spoke or made their discourse of protest, of denunciation and proclamation, through art, the sole manner of discourse they were permitted. (Freire 1994, p. 161)

Again there was the same pattern of education of the poor, political transformation, and folk or popular culture that existed in the work of Highlander and the Civil Rights movement.

The joining of popular culture into what developed as bottom-up models of social change are what made these movements unique. In Latin America, two interrelated social movements—the literacy approach of Paulo Freire and the Base Community movement of Liberation Theology—emerged between 1957 and 1985 to confront military power and economic injustice. In North America, the Civil Rights Movement in the 1950s and 1960s was based on a strategy of grassroots organizing that later led to collaborative research for social action (e.g., the work of Appalshop, and the later work of Highlander on issues, such as the rights of miners and environmental issues [Carawan and Carawan 1993]). Both cases (as well as the ideas of Grundtvig and Cabral, for that matter) linked common features: a triad of adult education/politics/popular culture; an educational strategy that incorporated life and field experience; methodologies that placed authority and analysis in the hands of the poor, particularly the rural poor; a sense that teaching as solidarity is ultimately an act of love and risk; and an ideal that the resulting transformation of social reality is the apex of the educational process. These models have differed from many other radical movements in two striking ways; they all questioned traditional hierarchical, top-down structures of education and political movements (even while there was a tension between that ideal and the realities of such issues as leadership and curriculum development), and they all depended upon a thorough understanding of the forms, customs, agricultural practices, land use, language, music, dance, and beliefs—in short, the folklife—of the communities in which they were situated.

The intellectual leaders of these movements were able to transmit (as well as develop) their ideas through small group gatherings, religious meetings, and other open discussions, as well as in educational institutions and informal settings, training a generation (or two) of cultural and community activists that shaped work for social justice in the Americas thereafter. They did so not by teaching abstract theory, but by teaching that theory could be synthesized from the lived experiences of oppressed people. And they did so, ideally, by teaching through a process that respected the idea of dialogue and promoted it as an egalitarian tool in the learning process. Still, for them, theory and practice were and are inseparable. Not just in the typical sense, that a theory is necessary in advance of a successful practice, but that, in Freire's words, "without practice there's no knowledge" (Horton and Freire 1990, p. 98), so the two are in a constant dialectic, and the synthesis is what Freire refers to as "praxis."

In the vein that runs from the Danish Folk Schools through the popular literacy campaigns in Central America there remains the ideal that the observation and analysis of social reality are at the core of pedagogy, whether teaching literacy, science, or ethics. Suffice it to say that for those pedagogues involved in this kind of work, there was, and is, a preference for observing social reality from the perspective of the poor, marginalized, and oppressed, a perspective usually ignored and overlooked in typical state curricula. Furthermore, reality as any of us live it is constructed by social forces, as Peter Berger and Thomas Luckmann famously point out (1966). Horton, Freire, and their students would be quick to point out that that reality has a validity (particularly when planning the transformation of the world into a more equitable habitat). It may not be the reality in an absolute sense, but it is a social perspective, no less real for not being absolute, which has been left out of state-sponsored schoolbooks for too long.

Perhaps it will be no surprise then that toward the end of their lives, Freire and Horton collaborated on a book, *We Make the Road by Walking* (1990), consisting entirely of extended dialogues between the two. The dialogic approach was not just representative of Freire's mode of teaching, it also was a practical consideration given Horton's frailty at the time, and indeed, Horton died within two weeks of his last meeting with Freire to review the manuscript. The book is more significant for showing the confluence of the two educators' ideas, rather than one's influence on the other, although we know that Horton knew of Freire's work at least as early as 1972 (Horton 1973). Both agree on the dialogic and dialectic nature of true education, where the teachers learn from the students, and that "believing in the people, but not in a naive way," as Freire says, is "necessary" (Horton and Freire 1990, p. 247). This leads to a bottom-up effect on the community organizer:

> [T]here is an interpretive and necessary moment in which the leaders who are trying to mobilize and organize have to know better what they are doing . . . They will change their language, their speech, the contents of their speech to the extent that in mobilizing the people they are learning from the people. And then the more they learn from the people the more they can mobilize. (Horton and Freire 1990, pp. 121–122)

Likewise, theory and knowledge are dialectical, and forever changing based on new information and historical circumstances. In Freire's words:

> I cannot fight for a freer society if at the same time I don't respect the knowledge of the people . . . Knowledge is changed to the extent that reality also

moves and changes. Then theory also does the same . . . . (Horton and Freire 1990, p. 101)

Freire's motivation is shaping social reality. Nowhere do we see more clearly than in this book that Horton, who also was raised within the culture in which he would later work, matches Freire in motive and technique.

One of the central points Horton makes in this dialogue, which speaks for the approach of both men, is that his school, his method, were not about reform:

> I think the poor and the people who can't read and write have a sense that without structural changes nothing is worth really getting excited about. They know much more clearly than intellectuals do that reforms don't reform. They don't change anything . . . Now if you could come to them with a radical idea . . . where they see something significant, they'd become citizens of the world . . .
> So to embolden people to act, the challenge has got to be a radical challenge. It can't be a little simplistic reform that reformers think will help them. It's got to be something that they know out of experience could possibly bring about a change. And we sell people like that short by assuming that they can take a little baby step and isn't this wonderful . . . But that kind of analysis doesn't fit the national situation in any way here in this country. So it leaves us working with the remnants, leaves us working with the little pockets of hope and adventurism wherever we can find it. (Horton and Freire 1990, pp. 93–94)

This is the crux of the issue for Horton, who wrote relatively little. Later on he leaves no doubt about the revolutionary nature of his concept of pedagogical process:

> We concluded that reform within the [educational] system *reinforced* the system, or was co-opted by the system. Reformers didn't change the system, they made it more palatable and justified it, made it more humane, more intelligent. We didn't want to make that contribution to the schooling system. (Horton and Freire 1990, p. 200)[15]

Therein lies the difference between what Olive Dame Campbell and Myles Horton took from the Danish Folk School system, and why Freire's technique of literacy training was such an important part of revolutions from Nicaragua to Guinea-Bissau and the community organizing that laid the groundwork for future revolutions. It was not just a question of making the system more functional or palatable, it was an opportunity to remake the entire system.

A Synthesis of Approaches

Comparing the educational philosophy behind the Danish Folk Schools and Highlander, on the one hand, and the popular literacy movements of Paulo Freire and in Central America and Guinea-Bissau and Cape Verde, on the other, makes some obvious similarities stand out. The participants were largely from peasant and rural classes. Educational praxis was experiential. Authority for knowledge and understanding of social reality rested in large part with the students; even when they didn't have certain skills, which could be taught, they had the insight and life experience. Music, dance, and artisanship were essential parts of the curriculum. And synonymous words were used to describe the overall education: folk/*Volk* in the Germanic-speaking countries, and *popular* in the Romance-speaking ones.

On a more subtle, and perhaps fundamental, level is the recognition in all these examples of the importance of language as an essential tool in the process of emancipation. Literacy, being able to read and write in your language, is without doubt a basic skill. But if that skill is not connected to using one's native language, or to thinking critically about social reality, then it is incompletely developed. In societies that are not completely oral, reading and writing are forms of thinking (Freire 1985, pp. 13–14). Beyond literacy, being able to speak, and to have control over one's words and one's own language, is part of an emancipatory education. Henry Giroux writes "The political nature of language is revealed in its use to structure and shape the communicative process" (Giroux 1981, p. 138). I would add that its *ability* to structure and shape the local culture renders language even more "political." In other words, whether asserting one's Danishness under a Germanic sphere of influence, asserting one's civil and human rights, or asserting one's liberation from colonialist socioeconomic structures, one has to be able to form and speak one's own words. Freire calls this "naming the world." It is a question of not being rendered silent and powerless, but being able to speak, speak out, or even sing out, to assert one's cultural, civil, and economic rights. Ownership of one's own language is a vital, and political, step. Freire concludes, "The fact is that language is inevitably one of the major preoccupations of a society which, liberating itself from colonialism and refusing to be drawn into neocolonialism, searches for its own re-creation. In the struggle to re-create a society, the reconquest by the people of their own word becomes a fundamental factor" (Freire 1978, p. 176).

The concept of orality was central to the work of Grundtvig and his followers, Horton, and Freire. While this clearly involves being able to formulate and give voice to one's own thoughts and sentiments (and those of the community, cultural, or class group affiliation), it also involves a mastery of the oral tradition. But mastery of the oral tradition is seldom a passive recitation of the words of others. Those who master the oral traditions know how to shape and reshape, form, create and re-create the words and ideas they have inherited and give them new meaning in new performance. Contrary to the orthodox academic belief that literacy by and large supersedes orality, the folk and popular school models presuppose that orality and literacy exist alongside one another as parallel and mutually necessary skills.[16] Freire is never more clear on this point than when he concludes one article: "The five-year program will eliminate illiteracy in São Tomé and Príncipe. The people will speak their word" (Freire 1981, p. 30).

In order to speak or sing from a position of empowerment, knowing one's traditions is practically a prerequisite—not merely to have an awareness of where one stands in society, in terms of class, historical background, ethnicity, and so on, but to have an appreciation for and indeed an ownership of one's own culture, customs, holidays, sacred beliefs, means of cultivation and sustenance. This goes beyond self-knowledge to mean instead a knowledge of where and how one relates culturally to the world, an understanding of the folk traditions, lifeways, work and occupational culture,[17] beliefs, ethics, and aesthetics (Freire 1998, p. 38), music, and physical movement that place us within a local culture and a larger nation-state.

If the basics in primary education for children are the accepted "reading, writing, and 'rithmetic," it would not be farfetched to say that under the adult pedagogies of folk and popular schools, the basics for adult education are reading, speaking, and folklife. This would indeed be a radical transformation of educational priorities, of course, away from mastery of mere skills, and toward a critical knowledge of how to employ linguistic and ethnographic techniques in the service of self-understanding and socioeconomic liberation.

Folklife and Democratic Participation

Since my academic training is as a folklorist, I would like to conclude my argument by looking through the lens my home discipline provides me. Folklife studies—which I refer to as a broader category than Folklore as a discipline—originated in the study of peasant tales, customs, and beliefs, and emphasizes craft and music as being as deeply laden with purpose and meaning as more materialist aspects of social life. Where the study of antiquities, later called "folklore," focused on oral and verbal genres in the eighteenth and nineteenth centuries, the larger rubric, folklife, included ways of living, surviving, and other customs of the peasant class. Folk literature was collected to assert independent national identities, and the study of folklore has been linked to Romantic nationalism. Folklife was documented also to establish a differential identity, but since food and housing were also studied, then issues of class moved to the fore. In general, folklife (known as ethnology in much of the world) became more widely practiced in subjugated nations, poor or isolated regions within nations, and nations bordering the major powers: Ireland, Wales, Scotland, Denmark, Sweden, Hungary, the Pennsylvania German and Appalachian regions of the United States, and southern Italy.

The significance of this context is twofold. First, in all of these places education outside of universities, not just collection, became an important feature of ethnological work; folk schools and open-air and indoor museums have democratically made knowledge available again to "the folk." Second, folklife added a second dimension beyond romantic nationalism—which could and has become its own instrument of exclusion and persecution (see Abrahams 1993)—that of democratization, the other side of Romanticism. In this regard, it is no accident that Freire frequently mentions beauty, aesthetics, and of course freedom, that Horton cites Percy Bysshe Shelley as a major influence (Horton 1990, pp. 29–31; Horton and Freire 1990, pp. 34–35), that American visitors to Denmark invoked John Keats, or that revolutionary songs or songs extolling human dignity became part of the curriculum in Denmark, Nicaragua, and elsewhere.

This emphasis on experiential education finds modern practical incarnation in the form of participant observation, the basic technique of ethnographic investigation. That we learn by doing fieldwork is true whether it is labor in the field or field research, and in fact, the Nicaraguan literacy workers were expected to keep field diaries, which themselves became pedagogical tools (Miller 1985). Participation in life rather than reliance on book knowledge is the source of knowledge in folklife as a discipline. Another important shared principle is that the authority rests with the people in the community, whether they are studying or being studied, organizing or being organized. Or to use Freire's terms, people are subjects in their own history, rather than objects of someone else's research (Freire 1985, p. 199). This is the hallmark of traditional folklife research, particularly in areas such as Ireland, Wales, and Scandinavia where we see the purest form of this intellectual paradigm.[18] It is this very acknowledgment of where authority lies that distinguishes folklife work in more modernized milieux, and I would argue what sets the folklife as a discipline apart from other approaches. This is what I maintain is most epistemologically revolutionary about the folklife method, whether it is taking place in folk schools or folklife departments. In a given social reality, the authority rests with those who experience, and often struggle against, that reality, rather than with those who study it from afar.

Lost, perhaps, is the idea Horton, Freire, and their followers made explicit–that the analysis of class and social reality are features not just of some revolutionary or social change movements, but of everyday educational settings as well. Such a pedagogy depends only partly on content, the lived experience of the oppressed. Content is important, but it is trumped by method. How do we, or any students of culture, observe, listen, analyze, and come to know? How can we do fieldwork except by dialogue? I would go even further to say that in the study and analysis of the reality of everyday life, which we best access through field research—by its very nature involving dialogue as well as reciprocity—there is potential for radical community transformation. We share a method that values trust and authority in people rather than in books, recognizes the importance of cultural expression, sends people out into the field, looks at society from the bottom up, and analyzes the bonds of community while reinforcing them. All of these techniques were shared by the "teachers" in the folk schools, popular literacy campaigns, and base communities, while ostensibly teaching language and literacy. If we as students go as far as our predecessors did in the observation and analysis of social reality, what about that third step, the transformation of reality, the practice that would in turn sharpen and extend our theory?

The challenge of studying Grundtvig, Horton, and Freire is that they implicitly and explicitly tell us that knowledge applied is the highest theoretical calling (while we who have been trained in universities have been conditioned to believe that such an ideal or outcome—i.e., that high theory would develop better from praxis rather than from disinterested research—is counterintuitive). Such humanistic values as listening and observing in the field, engaging in analysis through dialogue and discussion, and then developing an applied methodology that encourages us to utilize our research in solving social problems, may be the last pedagogical hope for constructive revolutions that are not based on ideals of hate and devastation.

NOTES

The author would like to extend special thanks to Dorothy Noyes and Margaret Mills for their continuous encouragement to contribute to this conference and volume, to Tom Ewing for insightful and helpful editorial suggestions, and to Margaret Randall, Gioconda Belli, David Azzolina, Nancy Gadbow, and Carlos Fernando Chamorro for research assistance and answering last-minute questions.

1. These elements in part constitute the hidden political history of folklore studies in America.
2. This reverential tone would be echoed, some 55–60 years later, in reports back from U.S. delegations visiting postrevolutionary Nicaragua. The historical parallel, however, was disrupted by the fact that the Reagan administration was far more threatened by Nicaragua and actively pursued its overthrow. Denmark, especially in comparison to the much larger Soviet Union, presented no such threat in the 1920s.
3. Whisnant makes the point that, to some extent, the folk school movement was so diverse that people tended to emphasize those aspects of its curriculum and ideology they were most sympathetic to.
4. This appears in an article for a general readership published in *Mountain Life and Work*, the magazine of the Southern Mountain Workers.
5. Or *fakelore*.
6. The Zilphia Horton collection in the Tennessee State Historical Archives contains over 1,300 song texts, including historical information, biographical notes on the composers, and the changes made at different points in the folk process. There are also 85 songs on tape, and 11 songbooks published at Highlander while she was there (Tennessee State Library and Archives 1967). The range of the songs varies from popular songs and composed standards ("As Time Goes By," "Anchors Aweigh," "The Star-Spangled Banner," "Old Folks at Home," and "On Wisconsin" are all in the collection) to a range of folk and labor songs, some widely-known and some topical. There are songs on race and injustice, such as "Black Man in Prison," "Ballad of the Blue Bell Jail," and "Scottsboro Boys Shall Not Die"; songs on politics, "The Guy that I Send to Congress," "There'll Always be a Congress," "Doing the Reactionary," "The Youth Act Must be Passed this Year," and "There is Mean Things Happening in this Land"; and songs on labor, such as "Bread and Roses," of course, as well as the less well-known "Workers' Marseillaise," "Ten Little Sweatshops," "Ten Little Textile Workers," "Song of the Danville Strikers and Their Children," "Battle Hymn of the Farmers' Union," "Ballad of the Southern Summer School," "Everett, November Fifth," "Fifty Thousand Lumberjacks," "Give Me that Textile Workers Union," "I'm Too Old to Be a Scab," "March of the Toilers," "Oh, Tortured and Broken," "Siamese Out

of Work Song," "Song of Local 102 (ILGWU)," "Song of the Munitions Workers," "Southern Tenant Farmers' Union Song," "The Steel Workers' Battle Hymn," "There Are Crowds That Make You Grumpy," "The Twenty-Third Shirt Factory Psalm," "Victory Song of the Dressmakers," "Victim of Priorities," "Waitresses' Song," "We Have Fed You all for a Thousand Years," "You'll Wish You Were One of Us," "You'll Wish You'd Listened to Us," and many other irresistible titles. Her collection also includes letters from Ben Botkin, Alan Lomax, Woody Guthrie, Tom Glazer and, as expected, Pete Seeger, and from representatives of such notable unions as the IWW, the ILGWU, the CIO, and the UAW (Tennessee State Library 1967).

7. In an odd historical quirk, the Nicaraguan literacy campaign began on March 24, 1980 (Chacón and Pozas 1980), the very day Archbishop Oscar Romero was assassinated in neighboring El Salvador. While coincidental, literacy and the forces of counterrevolution are linked in some cosmic or in this case metaphoric sense.

8. "Equity" as opposed to "equality," which connotes "sameness." Grundtvig even specifically chose—and coined (Begtrup et al. 1926, p. 82)—the Danish word *ligelighed* in contradistinction to *lighed*, which means "equality," for that reason. In one of his songs he wrote of "equal dignity in castle and cottage" (Bugge 2001, p. 73).

9. Horton first went to Nicaragua in 1980 (Horton 1990, pp. 202ff.) and Highlander sponsored a conference in Managua on North/South education in 1983 (Glen 1996, p. 275).

10. In Nicaragua, there had been two previous literacy campaigns with strikingly different political objectives. In the 1960s, under the Somoza dictatorship, one political party had launched a literacy campaign following Freire's techniques in the late 1960s (Berryman 1984, p. 59). A decade later, the Somoza government launched a literacy campaign as a tactic to infiltrate the popular oppositional movement and identify Sandinista sympathizers (Cardenal and Miller 1981, p. 4).

11. Whisnant does not address the literacy campaign as much as one would hope in his intensive study of the politics of culture in Nicaragua, *Rascally Signs in Sacred Places* (1995), perhaps because the literacy campaign was run from the Ministry of Education rather than the Ministry of Culture (as the two ministries were headed by brothers, Fernando and Ernesto Cardenal respectively, one could hardly imagine the programs could have been much closer). It would make an interesting point of comparison, since in his earlier work Whisnant faults Olive Dame Campbell's cultural and educational work in Appalachia for its avoidance of radical political ideology and social change (1983, pp. 135–136). Some of the Danish folk schools Campbell encountered were politically and explicitly radical, while she was more of a reformer than a revolutionary. Whisnant writes about what Campbell omitted from the Danes, but his own underemphasis of the literacy campaign in Nicaragua and, more surprisingly perhaps, his lack of mention of Horton and Highlander—they appear just in one endnote in *All that is Native and Fine* (1983, p. 299)—are curious.

12. See, notably, Freire (1978), but also Freire and Macedo (1987). Freire said, tantalizingly, "One of my dreams, which went unfulfilled, was to conduct a thorough analysis of Amilcar Cabral's work . . . In this book I would have drawn a clear distinction between 'revolutionary pedagogue' and 'pedagogue of the revolution.' We have some revolutionary pedagogues; but we don't have many pedagogues of the revolution. Amilcar is one of them" (Freire and Macedo 1987, p. 103). Cabral's—and Freire's—work was important not only in Guinea-Bissau

and Cape Verde, which shared a revolutionary movement but are now distinct nations, but also the much smaller island nation of São Tomé and Príncipe, further to the south but also colonized by Portugal, where the basic reading primers were called the *Popular Culture Notebooks* (Freire 1981, p. 28). As a point of comparison, as of the year 2000, these three countries together had a population of under two million; compared with Nicaragua, which had nearly five million, Denmark, which had about half a million more than that, El Salvador, which had over six million, and Cuba, which now has over eleven million people. This is a little deceptive for use in this paper, because Denmark's population growth was much slower in the twentieth century. A more useful comparison has to do with area, since transportation and diffusion of educational resources were probably more equivalent in nineteenth-century Denmark and twentieth-century Nicaragua. One realizes how all the more extraordinary the Nicaraguan literacy campaign was by observing that Nicaragua has three times the land mass of Denmark and around one-tenth the number of passenger vehicles. (For more information on the transportation difficulties, see Cardenal and Miller [1981, pp. 12–13].)

13. The bibliography on Liberation Theology is vast, but one of the best overviews of the movement in practice is Berryman (1984).

14. Ellipses between paragraphs represent questions posed, which have been edited out for reasons of space, or in one case an answer to a question not relevant here.

15. These are hard words to hear and take to heart during tough economic times. See also Coe (2000, p. 40) for similar concerns in a discussion of the Scandinavian folk schools.

16. The importance of orality in revolutionary situations was also demonstrated in the small contemporaneous 1979–83 revolution in Grenada, in which oral poetry, calypso, and the role of Creole all played a large part, even though the nation already had a very high rate of literacy (Searle 1983, pp. 92ff.).

17. Freire was more explicitly coming from the intellectual tradition of Antonio Gramsci, for whom self-knowledge in terms of class and history is critical: "The starting point of critical elaboration is the consciousness of what one really is, and is 'knowing thyself' as a product of the historical process to date which has deposited in you an infinity of traces, without leaving an inventory" (Gramsci 1971, p. 324). Although Gramsci was certainly interested in the question of rural intellectuals, he observed that the transition to industrialization, in eliminating skilled trades, also wiped out a layer of specialized knowledge that skilled laborers would have cultivated through the study of their trades. In schools run by the working class, intellectual studies and manual labors would be combined in the curriculum (Gramsci 1985, pp. 42–43), as indeed they were in the folk high schools in Denmark and at Highlander. Freire, citing Gramsci, also sees a false dichotomy between "muscular-nervous effort" and "intellectual-cerebral elaboration" (Freire 1985, pp. 71, 94 n. 13; Gramsci 1971, p. 9).

18. Nowhere more explicitly, for example, as in one of my favorite folklife book titles: *Ask the Fellows Who Cut the Hay* (Evans 1956). Or, for that matter, in the case of Brazil, in the revelations unearthed by the folklorist–protagonist in Jorge Amado's novel, *Tent of Miracles* (Amado 1971).

REFERENCES

Abrahams, Roger D. 1993. "Phantoms of Romantic Nationalism in Folkloristics." *Journal of American Folklore* vol. 106, pp. 3–37.

Adams, Frank with Myles Horton. 1975. *Unearthing Seeds of Fire: The Idea of Highlander*. Winston-Salem: John F. Blair.

Amado, Jorge. 1971. *Tent of Miracles*. Translated by Barbara Shelby. New York: Avon Books. [Orig. published 1970.]

Arnove, Robert F. 1986. *Education and Revolution in Nicaragua*. Westport, Conn.: Praeger.

Begtrup, Holger, Hans Lund, and Peter Manniche. 1926. *The Folk High Schools of Denmark and the Development of a Farming Community*. London: Oxford University Press.

Berger, Peter L. and Thomas Luckmann. 1966. *The Social Construction of Reality*. Garden City, N.Y.: Doubleday & Company.

Berryman, Phillip. 1984. *The Religious Roots of Rebellion*. Maryknoll, N.Y.: Orbis Books.

Bugge, K. E. 2001. *Folk High Schools in Bangladesh*. Translated by David Stoner. Odense: Odense University Press.

Campbell, John C. 1921. *The Southern Highlander and His Homeland*. New York: Russell Sage Foundation.

Campbell, Olive Dame. 1928. *The Danish Folk School*. New York: Macmillan Company.

Carawan, Guy and Candie Carawan. 1993. "Sowing on the Mountain: Nurturing Cultural Roots and Creativity for Community Change." In *Fighting Back in Appalachia: Traditions of Resistance and Change*. Stephen L. Fisher, ed. Philadelphia: Temple University Press, pp. 245–261.

Cardenal, Ernesto. 1982. *The Gospel in Solentiname*. Translated by Donald D. Walsh. Maryknoll, N.Y.: Orbis Books. [Orig. published 1975.]

Cardenal, Fernando. 1981. "Objetivos de la Cruzada Nacional de Alfabetización." In *Nicaragua: Triunfa en la Alfabetización: Documentos y Testimonios de la Cruzada Nacional de Alfabetización*. San José, Costa Rica: DEI and Ministry of Education. [Orig. published 1979.] pp. 27–45.

——— and Valerie Miller. 1981. "Nicaragua 1980: The Battle of the ABCs." *Harvard Educational Review* vol. 51, pp. 1–26.

Carter, Vicki K. 1994. "The Singing Heart of Highlander Folk School." *New Horizons in Adult Education* vol. 8, no. 2 (Spring), pp. 4–24.

Chacón, Alicia and Victor S. Pozas, eds., 1980. *Cruzada Nacional de Alfabetización: Nicaragua Libre 1980*. Managua: Ministry of Education.

Coe, Cati. 2000. "The Education of the Folk: Peasant Schools and Folklore Scholarship." *Journal of American Folklore* vol. 113, pp. 20–43.

Equipo DEI (Departamento Ecuménico de Investigaciones). 1981. *Nicaragua: Triunfa en la Alfabetización: Documentos y Testimonios de la Cruzada Nacional de Alfabetización*. San José, Costa Rica: DEI and Ministry of Education.

Evans, George Ewart. 1956. *Ask the Fellows Who Cut the Hay*. London: Faber and Faber.

Freire, Paulo. 1970. *Pedagogy of the Oppressed*. Repr. translated by Myra Bergman Ramos. New York: Seabury Press. [Orig. published 1968.]

———. 1978. *Pedagogy in Process: The Letters to Guinea-Bissau*. Translated by Carman St. John Hunter. New York: Seabury Press.

———. 1981. "The People Speak Their Word: Learning to Read and Write in São Tomé and Principe." Translated by Loretta Porto Slover. *Harvard Educational Review* vol. 51, pp. 27–30.

———. 1985. *The Politics of Education*. Translated by Donaldo Macedo. South Hadley, Mass.: Bergin & Garvey Publishers, Inc.

————. 1994. *Pedagogy of Hope*. Translated by Robert R. Barr. New York: Continuum.

————. 1996. *Letters to Cristina: Reflections on My Life and Work*. Translated by Donaldo Macedo, Quilda Macedo, and Alexandre Oliveira. New York: Routledge.

————. 1998. *Pedagogy of Freedom: Ethics, Democracy, and Civic Courage*. Translated by Patrick Clarke. Lanham, Md.: Rowman & Littlefield Publishers, Inc.

———— and Donaldo Macedo. 1987. *Literacy: Reading the Word and the World*. South Hadley, Mass.: Bergin & Garvey Publishers, Inc.

Giroux, Henry A. 1981. *Ideology, Culture, and the Process of Schooling*. Philadelphia: Temple University Press.

Glen, John M. 1996. *Highlander: No Ordinary School*. 2nd ed. Knoxville: University of Tennessee Press.

Gramsci, Antonio. 1971. *Selections from the Prison Notebooks*. Edited and translated by Quintin Hoare and Geoffrey Nowell Smith. New York: International Publishers.

————. 1985. *Selections from the Cultural Writings*. David Forgacs and Geoffrey Nowell-Smith ed. Translated by William Boelhower. Cambridge: Harvard University Press. [Orig. published 1920.]

Hammond, John L. 1998. *Fighting to Learn: Popular Education and Guerrilla War in El Salvador*. New Brunswick, N.J.: Rutgers University Press.

Hart, Joseph K. 1926. *Light from the North: The Danish Folk High Schools—Their Meanings for America*. New York: Henry Holt & Co.

Horton, Aimee Isgrig. 1989. *The Highlander Folk School: A History of Its Major Programs*. Brooklyn: Carlson.

Horton, Myles. 1938. "The Community Folk School." In *The Community School*. Samuel Everett, ed. New York: D. Appleton-Century Company, pp. 265–297.

————. 1944. "Grundtvig and Danish Folk Schools." *Mountain Life and Work* vol. 20, no. 1, pp. 23–25.

————. 1947. "Farm-Labor Unity." *Prophetic Religion* vol. 8, pp. 79–82, 93.

————. 1973. "Decision-Making Processes." In *Educational Reconstruction: Promise and Challenge*. Nobuo Shimahara, ed. Columbus: Charles E. Merrill Publishing Co., pp. 323–341.

————. 1983. "Influences on Highlander Research and Education Center, New Market, Tennessee, USA." In *Grundtvig's Ideas in North America*. Copenhagen: The Danish Institute, pp. 17–31

————. 1990. *The Long Haul*. With Judith Kohl and Herbert Kohl. New York: Anchor Books, Doubleday.

———— and Paulo Freire. 1990. *We Make the Road by Walking: Conversations on Education and Social Change*. Brenda Bell, John Gaventa, and John Peters, eds. Philadelphia: Temple University Press.

Knight, Edgar Wallace. 1927. *Among the Danes*. Chapel Hill: University of North Carolina Press.

Kozol, Jonathan. 1978. "A New Look at the Literacy Campaign in Cuba." *Harvard Educational Review* vol. 48, pp. 341–377.

Lewis, John, with Michael D'Orso. 1998. *Walking with the Wind*. New York: Simon & Schuster.

Manniche, Peter. 1939. *Denmark, A Social Laboratory*. Oxford: Oxford University Press.

Miller, Valerie. 1985. *Between Struggle and Hope: The Nicaraguan Literacy Crusade*. Boulder, Colo.: Westview Press.

Paredes, Américo. 1969. "Concepts About Folklore in Latin America and the United States." *Journal of the Folklore Institute* vol. 6, pp. 20–38.

Paulston, Rolland G. 1974. *Folk Schools in Social Change: A Partisan Guide to the International Literature.* Pittsburgh: University of Pittsburgh.

Rolando (pseudonym). 1985. Interviewed by author, in English, Philadelphia. May 29.

Rørdam, Thomas. 1980. *The Danish Folk High Schools.* Revised 2nd ed. Translated by Alison Borch-Johansen. Copenhagen: The Danish Institute.

Searle, Chris. 1983. *Grenada: The Struggle Against Destabilization.* London: Writers and Readers Publishing Cooperative Society Ltd.

Tennessee State Library and Archives, Manuscript Division. 1967. *Zilphia Horton, 1910–1956. Folk Music Collection, 1935–1956.* Nashville: Tennessee State Archives, Registers no. 6.

Tolstoy, Leo. 1862 [1967 reprint]. "On Popular Education." In *Tolstoy on Education.* Leo Wiener, Trans. Chicago: University of Chicago Press.

Whisnant, David. 1983. *All That is Native and Fine.* Chapel Hill: University of North Carolina Press.

———. 1995. *Rascally Signs in Sacred Places: The Politics of Culture in Nicaragua.* Chapel Hill: University of North Carolina Press.

6

Rite of Passage as a Communal Classroom

The Pedagogical Recycling of Traditional New Year Celebrations in Turkey

Yücel Demirer

You happen to be in Istanbul, or Ankara, or any major city in Turkey, on March 21. You are spending the day exploring and walking the lively streets. From a nearby park, you see a rising trail of smoke. Curious, you approach the park for a closer look and hear the sound of music. People of all ages are dancing together, women and men, hand in hand. Powerful chants and slogans permeate the air, sung with joy and determination. People jump through the smoke over a fire, with victory signs in their hands. You see a banner with "N-E-W-R-O-Z" printed on it. The colors red, yellow, and green are everywhere among the festivities, in different patterns and combinations. As you wander along, you notice the heavy security measures taken. Inconsistent with the joy, disturbing tension hangs in the air.

You then walk maybe just one mile down to the street, toward another group of people you have spotted. You are hearing music and seeing dancing again. This time there are more school children in their uniforms and with their teachers. There is a haircloth tent in the middle of this celebration, where people are cooking rice. There are fires and jumping again, but something is different. Here, security is less tight and there are fewer people but many Turkish flags. The banners spell "N-E-V-R-U-Z." The mood is somewhat different here. In spite of common characteristics of both celebrations such as dancing and singing, there is a different vibration in the air; the festivities feel more constrained, more official, and more constructed.

You leave the area and continue your walk, noting that "Nevruz" is emblazoned on the billboards and banners hanging between the traffic lights. Further along, you see two young people handing out yellow pamphlets that read "Newroz" again. The two words keep appearing, but never together in the same text. Is this a misprint or is there some meaning in the difference? More importantly, why would two almost similar festivals be held simultaneously, yet separately, and so close to one another?

This scene is the observance of the solar New Year, marked by the spring, or vernal, equinox. This pre-Islamic New Year celebration begins on the first day of spring, March 21. Although the origin of Nevruz/Newroz is vague, the foundation of the celebration goes back to Zoroastrian Iranian mythology, and is considered a pastoral festival celebrating the end of winter. Originally the word meant "new light," later coming to mean "new day" in Persian. Spellings transliterated from other regional languages include: Nawruz, Navroz, Nowruz, No Ruz, Nauroz, and Navrouz.[1] Traditionally, it has been a time for joy and peace throughout the region. In contrast to its common background, the festival has been appropriated and celebrated in various formats throughout the Middle East and Central Asia. While it was known as the official Farmers Day in pre-Taliban Afghanistan, it was also celebrated as an anti-Soviet resistance day in pre-Soviet Tajikistan. It has also worked as a Christmas-like visiting, gift-giving, and cooking day in Iran (Attar 1998; Glasse 2001, p. 342; Henderson and Thompson 1997, p. 290; MacDonald 1992, pp. 186–187; Shariati 1986; Thompson 1998, pp. 314–317). Today, two distinct and contested versions of this celebration are observed in Turkey: the Kurdish festival of Newroz and the official Turkish festival of Nevruz.[2]

This chapter examines the nature of these contested festivals and their pedagogical utilizations. It explores the development and function of contrasting, pedagogical activities through Nevruz and Newroz and draw out their comparative implications for the study of the Kurdish issue and its resolution in Turkey. I begin by presenting the conceptual framework underpinning this chapter. Subsequently, I outline the cultural policies of the Turkish Republic and analyze their impact on the Republic's nation-building efforts. The following section focuses on the Kurdish response. It looks at how the Kurdish political medium facilitates Newroz as a framework that permits a redefinition and promotion of ethnic identity, creating a reliable channel for communication and alternative civic education. I then describe the Turkish state's "tradition reclamation" operation through Nevruz. In particular, I argue that the success and effectiveness of Newroz within the Kurdish population in Turkey have inevitably heightened the promotion of an official version of Nevruz and its pedagogical application. The concluding section discusses some of the broader implications of the Newroz/Nevruz contestation in Turkey. It suggests that even though the competition between the Kurdish Newroz and the Turkish Nevruz is widely seen as a source of political instability, the common characteristics of Nevruz and Newroz highlight their potential for conflict transformation and resolution.

This chapter looks at the development and execution of these contested versions of the solar New Year celebration and considers unique reappropriations of the "old" cultural sources intended to establish (and/or protect) identities required (and/or challenged) by "new" sociopolitical circumstances. These efforts are, in Stuart Halls's words, part of "never completed—always in process" self identification operations that are "constructed on the back of a recognition of some common origin or shared characteristics with another person or group, or with an ideal" (Hall 1996, p. 2).

This chapter also demonstrates the central role that Newroz celebrations have come to play in shaping and reproducing the Kurdish identity. In the Turkish republic, historically the Kurds could not freely use and disseminate Kurdish language and literature for fear that it would strengthen their sense of nationalism. The Newroz festivities, therefore, became a convenient way for Kurdish political leadership to circumvent those restrictions and craft a collective response toward the Turkish political system. This work seeks to investigate the processes by which these contested celebrations are structured and utilized. The main focus, therefore, is not to compare the two versions of the same new year celebration per se, but to compare how the celebration is used within the Kurdish "civic education," on the one hand, and the Turkish "official pedagogy," on the other.

My approach to this topic assumes that a sense of identity is a "conscious awareness" (De Vos and Romanucci-Ross 1995, p. 367). It is created through multilayered negotiations in various levels and provides mechanisms of understanding of self and the rest of the universe. This "conscious awareness," with many other factors, is shaped by highly tailored efforts of systematic socialization practices that are attached to pedagogical procedures.[3] This chapter thus examines these celebrations as a form of pedagogy, or what Peter Murrell refers to as "knowledge in practice," in terms of their relationship with the surrounding political climate and ideological constructions (Murrell, Jr. 2002, p. 60).

Consistent with the weight of political power in pedagogical decision-making processes, throughout the history of the Turkish republic the teaching and learning settings functioned as microcosm of the sociopolitical order and the citizens of the state were depicted as Turkish. The distinction of other identities was ignored in the pedagogical texts. Therefore, the Kurdish political organizations created alternative patterns and venues for learning and unlearning, which constitute a kind of "civic education." This practice clearly represents a different way of socialization and has to be distinguished from the common definition of civic education as political intervention that shapes the citizens' understanding and the level of participation in the governance by providing some ideals and standards (Torney-Purta et al. 1999). In the case of Kurdish minority of Turkey, I use the concept in a much narrower sense.[4] Here, "the Kurdish civic education" concept defines a less structured, moderate and condensed, but ultimately politically charged, effort to respond to the dominant cultural policies and "the disjuncture between curricular and social change" in Turkey (Anderson and Landman 2003, p. 7).[5] This chapter thus explains why Newroz was utilized by the Kurdish political leadership and how the success of Kurdish Newroz resulted in the reclaiming of the tradition by the cultural agencies of the Turkish state.

From the beginning of the Turkish Republic, cultural sources were highly utilized as an extension of the political agendas by the state as well as its various oppositions. These two festivals are taken into account as part of this heritage, therefore, as venues of cultural production as well as competition. In the context of the Kurdish narration of ethnic identity, Newroz celebrations have played a major role in shaping Kurds' collective response toward the Turkish political framework. Although it has been a part of the cultural repertoire of many peoples in the Middle East, including the Turks and the Kurds, it was only after the 1970s that the festival became a significant collective venue for the assertion of a distinct Kurdish identity. For the reason that the Kurdish national movement has successfully utilized this festival to differentiate and underline its identity, the Turkish state began in the 1990s to officially promote the Turkish version of the same celebration in an effort to eliminate and undermine the Kurdish identity claim. This is the process that I consider the recycling of a cultural celebration. The Turkish Ministry of Culture records show that although Nevruz was a part of Turkish culture, it was not a major celebration in the official celebration repertoire in the republic's earlier years. It was only in the 1990s, after observing the popularity of Newroz within its Kurdish population, that the Turkish state recycled its own version to override the Kurdish version and maintain the Turkish national identity oriented political order.

The recycling of a tradition or celebration does not merely imply a salvaging, but also a deliberate recontextualization of cultural objects toward a new purpose at the cultural level. The major function of recycling in the case of Nevruz is the rearrangement of the cultural zone to respond to a current threat by reusing an item that is part of the cultural depository but has not been in official use for a long time.[6]

RITUALS OF NATION BUILDING

How, and under what specific circumstances, did these two distinct pedagogical recyclings of the traditional Newroz and Nevruz celebrations take place? As a result of systematic efforts to organize Muslim people against the occupation, the independence struggle led by Mustafa Kemal Atatürk was highly supported by the Kurds of *Anatolia*[7] (Bruinessen 1992, pp. 272–273; Kirişçi and Winrow 1997, p. 75; Perinçek 1999, pp. 111–129, 197–212). Because of the war-related mortality, epidemics, large-scale migration and postwar population changes at the end of Turkey's War of Independence, the non-Muslim population of *Anatolia* had declined. Two large Muslim groups, Turks and Kurds, constituted 98 percent of the land's population within the boundaries of the Turkish Republic (Zürcher 1998, pp. 170–172).[8] The Turkish ethnic framework used to make the Turkish republic was defined in these terms in October 1919 by Kemal Atatürk:

> Gentlemen, this border is not a line which has been drawn according to military considerations. It is a national border. It has been established as a national border. Within this border there is only one nation which is representative of Islam. Within this border, there are Turks, Circassians, and other Islamic elements.

Thus this border is a national boundary of all those who live together totally blended and are for all intents and purpose made up of fraternal communities. (Quoted in Ahmad 2003, p. 80)[9]

After the success of an arduous struggle and a series of wars, however, the newborn republic was based solely upon the Turkish national and cultural identity. The search for national identity within the newly established Turkish republic followed a one-dimensional path succinctly summarized in *The Economist*'s "Survey of Turkey" published in 1996: "Turkey is more like a tree, with roots in many different cultures and ethnicities. In its early years it was pruned and trained to grow strictly in one direction: Turkish. Now, in its maturity, its branches tend to go their own way, seeking their own kind of light" (quoted in Nachmani 2003, p. 33).

This explanation, of course, addresses only one aspect of the equation.[10] Internal factions and lack of a unified voice among the Kurdish political elite also contributed to their exclusion. After the Young Turk Revolution of 1908, during the relatively promising political climate of the period for the Kurds in the late Ottoman era, a number of political clubs and schools founded by intellectuals, army officers, and some *aghas*, used Kurdish as their language of education.[11] The first series of Kurdish newspapers was also published during this time. However, these activities did not lead to a growth of a unified political movement. The Kurdish social structure continued to be dominated by tribal loyalty, in which the traditional elite was concerned more with the continuance of their local power than with making wider coalitions with other factions of the Kurdish leadership.[12] The *aghas*' mistrust of the urban Kurdish intellectuals and their campaigns to raise literacy and political consciousness endangered their position; this also hindered attempts at a unified voice and identity (McDowall 1992, pp. 30–31). These factors, along with the Western discourse of modernization, produced an image of Kurdish social structure as backward and feudal. Thus, the political antagonism between the Turkish national project and Kurdish nationalism could easily be presented as a clash between secular Turkish modernism and "tribal, backward, custom bound, ignorant, fanatically religious" tendencies of the Kurds (Bruinessen 1992, p. 274; Houston 2001, p. 89; Yeğen 1996, p. 216).

The construction of postwar normality came with both reforms and connected problems. Consistent with the dominant political developments of the post–World War I era and as a reaction to the deficiency of the multinational Ottoman political and administrative heritage, the founding fathers of the Turkish republic imagined a nation state (Anderson 1993; Gellner 1983, 1994; Hobsbawm 1992; Hobsbawm and Ranger 1983). Highly tailored programs to create "the monopolization of the means of symbolic reproduction and the monopolization of the means of physical force" served the goal of creating a homogenous nation (Jung 2001, p. 69)[13] This preference, predictably, brought the sense of exclusion for other ethnic identities. The resulting processes of nation building produced still unresolved conflicts and these were followed by decades long series of rebellions and forced displacements in Turkey.

By the 1970s the Kurdish "ethnic quest for identity" (Fernandez 1986a) surfaced in a new format, Newroz, to develop a Kurdish historical consciousness and sense of identity.[14] As a response to the state policies, Kurdish political leadership, then mostly young educated urban activists, reappropriated and promoted Newroz as a Kurdish ethnic narrative.[15] Consistent with the surrounding conditions that affected cultural production, the Kurdish cultural awakening took the stage as a revival of this tradition. Thus, Newroz became a powerful venue distinguishing the Kurdish minority from the dominant Turkish identity. These became, in Barbara Myerhoff's words, "moments of teaching, when the society seeks to make the individual most fully its own, weaving group values and understandings into the private psyche so that internally provided individual motivation replaces external controls" (Myerhoff 1982, p. 112). This move had dual functions. While on the one hand it was appropriated to respond the official cultural hegemony, it also became a widely used practical instrument to revitalize the political opposition at the popular level.

Both Newroz and the Nevruz celebrations reappropriate an available traditional source to legitimate and nourish two different projects for the future.[16] As could be expected, the pedagogical sphere corresponds to these two contrary stances and the alternative performances produce formulas for the issues under contention. Nevruz nurtures the social and political realm established by the Turkish state. Newroz, on the other hand, creates a powerful alternative narrative that hosts various manifestations of Kurdish identity. This move not only expands the limits of actorship toward the nonstate actors, but it also challenges the monopoly of the official pedagogy.

My initial encounters with these counteradaptations of a single festival raise a series of questions to be explored in this chapter. What are the institutional patterns of these pedagogical campaigns? In the limited venues for self-articulation, and in the midst of strict sociopolitical conditions, what does the pedagogical recycling of a one-day event tell us? Faced with debilitating factors that hinder research in the Republic of Turkey, what does this case of pedagogical recycling provide to scholars interested in investigating the "Kurdish question" of Turkey?

Since the "Kurdish question" has been one of the most sensitive topics in Turkey, the process of research was restrained. In the current Turkish political, social, and intellectual environment, and with the limited variety of conventional written sources, I intended to use multiple venues to overcome the difficult circumstances. To investigate the identity creation through celebrations and their pedagogical applications, I employed an ethnographic approach in both the gathering and interpretation of the data. In 2001 and 2002, I visited the Newroz festival sites and the archive of the Turkish Ministry of Culture and the Library of the Turkish National Assembly. I utilized the celebrations as a means of identifying the ethnic/national discourses. In my archival research, I focused on official documents to piece together the history and the day-to-day conduct of the official Nevruz.

My research findings established the Kemalist Turkish revolution as a continuous operation. It outlines how the Turkish nation-building experience repeats and reproduces itself in the face of new situations. It also enlightens the contrasting adaptations of Nevruz/Newroz for alternative civic educations and their potential for conflict resolution in Kurdish–Turkish relations. The findings suggests that even though the competition between Newroz and Nevruz highlights the differences between the two parties and is widely seen as a source of political instability, the commonality between the two theses surfaces during these events and raises hopes for resolving the protracted conflict.

The Kemalist Cultural Revolution

The pedagogical recycling of the New Year celebration is a fairly recent phenomenon in Turkey. In the Kurdish case, the festival's revival begins with the above-mentioned ethnic and political awareness movement in the 1970s; in the Turkish case, the festival was officially defined and encouraged by the state in the 1990s. The origins of these recent developments, however, date back to the founding of the Turkish republic in 1923, and can only be understood by considering the history of the republic's early cultural reforms and the mindset of its leadership. Considering the long-term effects of Turkey's cultural revolution, I need to discuss the major characteristics and the background of this "continuous" revolution.[17]

My main goal in this section is to illustrate the roots and the residue of the Kemalist revolution within the Turkish state tradition. As is seen, I argue that instead of being a one-time event, the Turkish cultural revolution situated itself as ongoing shaping of national consciousness, and constructed the necessary agencies accordingly. Influenced by Western thought, Kemal Atatürk initiated the transformation of his state on the ideals of modernization and development (Kili 2003, pp. 156–165, 222–227). However, Kemal Atatürk did not develop a comprehensive ideology that determined the transformations of Turkey. In Suna Kili's words, Kemalist ideology "had not become sufficiently systematized during Atatürk's lifetime":

> It is an ideology that grew out of action—out of the period of the Turkish National Struggle and out of the Atatürk reform movement. At times, it was the ideas that prompted action and at other times, it was the political and historical events that gave birth to an idea. (Kili 2003, p. 288)

Therefore, this tendency rooted a pragmatic approach in the macro decision-making processes, responding to conflicts as they arose. This pragmatic approach consequently became a part of the administrative traditions and formed the base of legitimacy for future administrative operations.

Kemal Atatürk, the founding leader of the Turkish republic, was born in 1881 in Salonica. He was educated in military schools. As a cadet he learned French and became acquainted with liberal ideas. The Ottoman poet of liberty, Namık Kemal became his main source of youthful inspiration. He was influenced by Western philosophy and was particularly attracted to

Positivism as a crucial framework for dealing with the causes of the decline of the Ottoman regime. He believed that education was the engine of progress and impediments to the creation of universal civilization had to be removed (Ahmad 2003, pp. 75–76; Mango 2002b, pp. 9–16).

Later on, during his volunteer service in Libya, he personally observed the powerlessness of the idea of Islamic brotherhood by facing revolts of the Arab independence movements against the Ottoman State. While serving there, he not only lost his "romantic, patriotic optimism, characteristic of his genera-tion of officers," but also his hopes for the continuance of Ottoman multi-nationalism (Mango 2000, pp. 103–108). The Balkan Wars and the following 1913 treaty that ended the war brought another reason to support his lack of hope for the future of the Ottoman regime. As part of the mass population exchange, his mother and sister had to leave Salonika and settle in Istanbul. This added a significant personal dimension to the military and political defeat that the collapsing Ottoman state was facing (Kinross 1965, p. 65; Mango 2000, p. 122). When the Ottoman war ministry and the military leadership were starting to be subordinated by German officers in 1914 in order to carry out reforms that were considered necessary after the Balkan Wars, the power struggle was observed cautiously by a group of military lead-ers, including Kemal Atatürk (Mango 2000, pp. 124–127). This, along with other factors, estranged Kemal Atatürk from the idea of the continuity of the Ottoman reign.

The Turkish national rebirth took place after the end of the Ottoman rule. The Ottoman State's defeat in World War I was followed by a political and military resistance led by Kemal Atatürk and his close circle of comrades. After the victorious end of the war of independence, the Turkish republic was founded on October 29, 1923. The newly established regime launched sys-tematic reforms to exceed the Western framework for the creation of the new state and society. The main principles of Kemalism became the road map for the development of the newborn republic: Republicanism, laicism, national-ism, populism, statism, and revolutionism (Gellner 1994, pp. 81–91; Gülalp 1997, pp. 54–56; Jung 2001, pp. 59–62).

In the ten years between 1913 and 1923, Anatolia's population and socioeconomic structure had dramatically changed. The cities became less habitable because of war, and the emigration of the non-Muslim populace left them devoid of most of the entrepreneurs and skilled labor, thus deplet-ing Anatolia of a promising economic base (Zürcher 1998, p. 172). Under these circumstances, in the words of Adam Lebor, "as arch-westernizer, he [Kemal Atatürk] was determined to drag Turkey into the twentieth century" (Lebor 1997, p. 222). In 1923, Ankara, then a small, unknown Anatolian town, became the capital of Turkey and the base for momentous social and political change.

To complete the cultural component of this political achievement, the leadership of the new republic intended to change the public sphere radi-cally.[18] In 1923, the Republican People's Party, then the leading single party, was formed, the treaty of Lausanne that secured the international status of

Turkey was ratified, and the Turkish republic was proclaimed. The caliphate, the potential rival to the republican leadership, was abolished and the last caliph exiled. The *medreses* and *mektebs*—religious schools—were closed and all schools put under the Ministry of Education. The Ministry of *Seriat* (Islamic Law) and *Evkaf* (Islamic Charity) was terminated. The courts in which the Islamic Law was administered were abolished. Religious endowments came to be controlled by an office directly under the prime minister. Religious endowments came to be controlled by a newly formed Ministry of Religious Affairs. The banning of the *fez*—a cylindrical red hat—and similar cultural measures carried the campaign into a new phase. The Gregorian calendar was accepted. The entire legal system of the country was renewed and translations of the laws from various European countries came into use. The Latin alphabet was adopted and the public use of Arabic alphabet was prohibited. Polygamy and the veil were outlawed; by 1930, women had the right to vote (Davison 1998, pp. 147–165; Lewis 1968, pp. 260–279). Parallel to this campaign, and consistent with its goals, was a new set of official or semi-official agencies that was created to implement the new cultural policies.[19]

Within the framework of the Turkish revolution, the locus of national identity shifted from Ottoman-Islamic culture to the pre-Islamic Asiatic culture of the Turks.[20] In the late Ottoman era, the word "Turk" had negative connotations and defined the unsophisticated peasant or dweller (Ahmad 1993, p. 78, 2003, p. x).[21] To deal with this stereotype and to create a new sense of national identity that was distanced from the Ottoman multinationalism and the idea of Islamic brotherhood, the Turkish History thesis was created. In Stanford Shaw's words, this thesis was created "to show the Turks what they have done in history" and used as "weapons to achieve the Republic's aims" (Shaw and Shaw 1977, p. 376). This thesis, first launched at the First History Congress in 1932, was a junction of the reaction against perceived European views of the Turks as inferior and the need for nation building. It asserted that the Turkish nation was a part of the history of civilization prior to entering to the Islamic world. This official thesis emphasized both the Asian-Turkic roots as well as the early civilizations of Anatolia (Mardin 1981, p. 211; Poulton 1997, pp. 101–109).[22]

Through this move, the new Turkish republic attached itself to the nationalist values of the Western world and distanced itself from Islamic civilization. Besides systematically repudiating the Ottoman past, the Kemalist project created and employed new sources for the Turkish nation-building process and the transformation of a traditional society. These actions required the creation and promotion of a collective national identity. During the implementation of these processes, civic education became central to the creation of the nation project and widely deployed cultural items, especially the traditional celebrations. However, the Turkish modernization/Westernization project did not create a political and administrative system based on well-defined rights of citizenship (Owen 1997, p. 249). This situation left the door open for future radical cultural and pedagogical policy-making and implementation both by and against the Turkish State.

THE KURDISH RESPONSE: THE PEDAGOGY
OF EMPOWERMENT

In the aftermath of the World War I, the treaties of Sevres (1920) and Lausanne (1923) shaped the political future of the region. According to the Treaty of Sevres, an international commission would prepare the Kurdish dominated region for its autonomy, and if the Council of the League of Nations agreed, the Kurdish population of the designated areas would have the right to vote for independence (Kurubaş 2004, pp. 95–102; McDowall 2000, pp. 136–137, 464–465).[23] However, the Treaty of Lausanne, signed after the successful war of independence, took an extremely firm stand against the terms of the Treaty of Sevres. It did not include the rights of the Kurdish minority and presented a very different future for the Kurdish people. The Turkish republic's sovereignty was established, and the redrawing of national borders divided the Kurds into five different states (Aydın 2002, p. 91).[24]

The Treaty of Lausanne was thus the first blow to Kurdish nationalism. The second blow came in 1924, when the new constitution did not have the same stipulations for Kurdish local rights granted in the previous 1921 constitution.[25] Leading up to the foundation of the Turkish republic Kemal Ataturk preferred to use the term "Nation of Turkey," though he employed the term "Turkish Nation" in his later speeches (Oran 1996, pp. 37–38). An official history based upon Turkish identity thus refused to acknowledge the existence of Kurdish ethnicity in Turkey. Since then, the Turkish founding fathers' ideology, focusing on modernity and Westernization, opened the way for the agents of Turkish nationalism to freely utilize social and cultural policies to respond to all sorts of oppositions (Chailand 1993, pp. 73–77; Fuller 1999, pp. 232–233).

The establishment of this national framework was followed by a series of rebellions organized by the Kurdish oppositionists.[26] The first major wave of these revolts was harshly repressed and ended in 1938. Thousands of Kurdish citizens of Turkey were forced to emigrate in various parts of the country (Bruinessen 1994, pp. 144–154; Chailand 1994, pp. 36–39; Kabacalı 1991; O'Ballance 1973, pp. 27–29). Further responses by the Turkish republic included closing religious organizations thought significant to the organization of these rebellions. Another important outcome that shaped the general understanding of the Kurdish issue in Turkish society was the gradual portrayal of any opposition to the national agenda as a movement of traitors bent on collapsing society rather then establishing their own due rights (Olson 1989, pp. 159–160).

After World War II, a relatively liberal political climate emerged in Turkey. The new political environment of the 1950s eased assimilation practices (Bruinessen 2000, p. 227). Although the military coup in 1960 revived anew the Turkish government's heavy handed policies toward Kurds, the 1961 constitution created a more liberal environment embraced by the Kurdish opposition. The patron client nature of the Turkish political system saw several traditional Kurdish leaders from mainstream parties being elected to the National Assembly. In another major shift on the Turkish political scene, several candidates from the popular Marxist party, Turkish Workers Party, won parliamentary seats in the 1965 elections (Bozarslan 1992, pp. 97–98). The resolution of the fourth congress of the Turkish Workers Party on the Kurdish issue became critical to push the Kurdish minority issue to the forefront of mainstream politics (Lipovsky 1992, p. 78).

Another boost to the increase of the Kurdish opposition occurred in late 1969 when major worker demonstrations took place in the streets of Istanbul, the largest city of Turkey. Students of Kurdish origin, affiliated with the DDKO (Eastern Revolutionary Cultural Centers) were also turning to radical methods to raise awareness for their nationalistic cause. The cultural centers soon expanded, weaving a network across many Kurdish towns and urban centers of Turkey (Bozarslan 1992, pp. 99–101).

The work of these centers apparently resonated with young, educated Kurds in major cities, because many became active in political parties and organizations that were associated with the Turkish left. This growing political activism among educated Kurds had a significant impact on the Kurdish population of Turkey, as it brought the Kurdish issue to the forefront of the political scene and planted the seeds for the growth of the opposition currently seen in Turkey today (Bruinessen 1990, pp. 44–45).

Under these conditions, the symbols of Newroz started to appear in various political meetings.[27] Newroz came to be seen and utilized as a "situated activity system" for various sorts of encounters, including pedagogical operations (Goffman 1961, p. 8). Especially in the 1980s, Newroz celebrations began to play an increasingly important role in shaping Kurds' collective response toward the Turkish political system. As was the case with many invented processes of identities, it was the Kurdish political elite that selected Newroz to synthesize and systematize the Kurdish ethnic identity. These young activists incorporated elements of the traditional culture with elements from the current political culture, thus using folklore material for political purposes. It was from here on that Newroz began to carry manifestations of the Kurdish identity and was used as a political and pedagogical instrument by Kurdish political leadership. Throughout the 1970s, symbols of Newroz were visible in political demonstrations. It was also during this time that Newroz was officially linked with separatist terrorists and denigrated as troublesome. By the 1980s, the Newroz tradition had become intimately linked with the Kurdish political agenda in Turkey.

Symbols of Newroz have also marked important historical milestones throughout the history of the Kurdistan Workers Party (PKK)[28] that spearheaded the guerilla wars against the Turkish state.

The eminent existence of Newroz can also be seen throughout the history of the PKK.[29] The PKK led the guerilla war against the Turkish state. It was established on November 27, 1978, under the leadership of Abdullah Öcalan. Its objectives and the organization structure have varied throughout its history (Güneş-Ayata and Ayata 1999, pp. 133–134; Gunter 2004, p. 119;

White 2000, p. 135). Mazlum Doğan's, one of the leaders of PKK, timing for committing suicide on March 21, 1982, to protest the prison conditions in Diyarbakır Prison No. 5, stands as a significant example of Newroz's place within the party vocabulary.[30] On Newroz day in 1990, Zekiye Alkan, a medical student in Diyarbakır threw herself into a fire after saying, "from now on, it's time to stoke the Newroz fire not with twigs but with human flesh" (Polat 1991, pp. 65–66; Zana 1991, p. 325). Gunter also notes that during its long running rebellion between 1978 and 1999, the PKK often started its spring offenses on Newroz (Gunter 2004, pp. 147–148).

These extreme reinterpretations of the tradition unite the cultural aspects of the Kurdish experience with the learning procedure that is attached to the political struggle. The timing of these suicides with the Newroz celebrations, as well as their use of fire, which has a central position in the festivities, illustrates the application of the historical–cultural framework as knowledge in practice.

Despite these dark extensions of the festival, Newroz is ultimately a celebration. The format of Newroz in the Kurdish context has been a one-day public event celebrated on March 21. Although it is celebrated in various urban and rural settings, the major gathering takes place in Diyarbakır. During the celebration, men and women dance and sing Kurdish traditional and modern songs. Huge bonfires are set on the meeting ground and people jump over the fires. Music and political speeches are a crucial part of the celebration. Community and political leaders adapt the general Newroz theme to address the present political situation. Banners present the current agenda of the Kurdish leadership.

During the 2001 and 2002 celebrations in Diyarbakır, for instance, opposition to the death penalty was the main theme, as proclaimed on the banners and slogans. (At the time the death penalty was a possibility for imprisoned PKK leader Abdullah Öcalan.) In 2003, the major theme was summarized on a large banner in the Diyarbakır Newroz site: "Neither denial nor extermination, [but a] democratic republic."[31] This paraphrases a struggle against the denial of Kurdish identity and the dissolution of their culture, while also making a demand for political inclusion.

As a grand public celebration, Newroz highlights and differentiates the crucial components of Kurdish identity. The festival displays and recalls the historical struggle between the Kurdish people and its oppressors. By transmitting the knowledge and the consciousness of the imaginary collective past, it becomes a medium to teach history and create a common culture. At the site of Newroz in Diyarbakır, for instance, carefully presented details bring together both the institutionalization of Kurdish culture and current political protest.

In the Newroz festival site, from the pro-Kurdish political slogans to the widespread use of the red–green–yellow colors[32] on clothes and banners, which cannot even be imagined in an ordinary Diyarbakır day, the dominant Turkish political discourse is momentarily displaced. The carefully chosen and arranged mythicized past becomes real (Geldern 1993, pp. 43–44). In Barbara A. Babcock's words, Newroz enables participants to free themselves from the limitations of "thou shalt not's" and enables them to make counter statements about the present social order. In this context of reversal, the "inverted beings" of the Newroz festival question the order in which they live (Babcock 1978, pp. 21–28). The festival site also serves as a public stage for people to perform. Although during the recent years Newroz celebrations are allowed and legally celebrated, being at the celebration site still carries risks.[33] Thus, simply being there becomes a major part of the festival participation. Parallel to this, participation in the celebration plays a crucial role in strengthening the communal ties. Moreover, the meeting ground functions as a medium for civic education and allows people to expand the limits of official pedagogical lines.

How, and why, has Newroz become a crucial venue for Kurdish civic education of Kurds in Turkey? The Kurdish search to withstand the Turkish cultural hegemony in the absence of venues for self-articulation led to the rediscovery of this ancient cultural celebration. As a response to the limiting social and political conditions, Newroz was put to use not only for claiming the Kurds' distinctive history, but also as a major medium for Kurdish civic education and for shaping their collective response to the political power.

Because of the limitations of their language and literary production, Kurds had to employ a fresh and unconventional source for civic education. Lack of suitable conventional sources for this purpose made them create an "experience-near" concept from an "experience-distant" frame (Geertz 1984, p. 124). As a result, the conventional direction of the learning process, from elder to younger, had radically changed. In the sociopolitical rebirth of the Kurdish Newroz, an unusual way of learning—what Margaret Mead called "prefigurative learning"—had arisen in its stead, and the flow of the direction of learning was reversed (Mead 1978). This change in the directional flow of learning—from the younger to the older—was first noted by Yılmaz Varol (1994, pp. 47–48).[34] Varol, in his search for the origins of Newroz, argued that while myths and legends are usually recounted by the elders, in the case of Newroz, it is the younger generation who became the teachers of such knowledge to the rest of the Kurdish society.

During my two fieldtrips to Diyarbakır in 2001 and 2002, when I asked about the beginning of the Newroz tradition, I was always told that Newroz has been with the Kurdish people forever. However when I asked specific questions such as "When did you first hear of such a thing?" or "When was the first time Newroz was celebrated in your village/neighborhood?" the responses began to be blurred. The response that I received from an old man during the 2001 Diyarbakır Newroz summarizes the impression that I had throughout my fieldtrips. When asked, the man replied thus:

> We have always had Newroz, but it was not like this [gesturing toward the festival ground]. Lately, youngsters told us the whole story. They told us the struggle of our people against merciless Dehak led by blacksmith Kawa and so on.

Such a response reflects fusion between the imaginary past and the present circumstances that situates a person on the social, historical and political map. It also informs us of multiple inversions between old/young, teacher/learner, and dominant/recessive. This inversion involves numerous implications for the construction of the Kurdish identity. As I mentioned earlier, the Kurdish revival expanded from the political sphere through a series of culture based moves, including Newroz. Politically charged community education became one of the main targets of this trend. As Mead has argued in *Culture and Commitment*, cultural transmission can occur in various ways in different types of societies. In order to differentiate the passing of culture from one generation to another, Mead suggests the terms *postfigurative, cofigurative,* and *prefigurative* (Mead 1978, p. 13).

Mead uses the term postfigurative when the future repeats the past. Cofigurative describes the circumstances in which the present becomes the guide for future contemplation. Prefigurative learning, on the other hand, occurs in dynamic societies where rapid changes take place and the elder generation is caught unprepared. Elders/teachers find themselves having been educated by the young generations on experiences that they themselves have not had. In this kind of context, the elders cannot provide the necessary knowledge to deal with the present situation (Mead 1978, p. 73). Although Mead offers these concepts to determine the models of cultural transformations on the global scale and mainly underlines the importance of related technological developments, her concept of prefigurative learning helps to make sense of the Kurdish way of learning through Newroz. It helps us to understand how various generations jointly take part in the construction of the Kurdish collective identity.

Newroz's pedagogy of empowerment has also changed "the culture of silence" (Freire 1985, p. 75). Stretching and enriching the limits of the Kurds' identity struggle, it brought new actors to the social and political scene. The ground of Newroz has become a space for an internal dialogue and the reproduction of the Kurdish cultural identity. It emerged as a significant challenge to the dominant patterns of social, political, and cultural oppression and arose in opposition to the Turkish official education.

THE TURKISH STATE'S RESPONSE: THE PEDAGOGY OF DOMESTICATION

At first the Turkish state had ignored the Nevruz/Newroz tradition and did not give it a significant role in its official or semiofficial list of celebrations. After observing the popularity and the wide acceptance of Newroz among the Kurdish minority in 1980s and 1990s, however, the Turkish state sought to reclaim this already invented tradition. The "dynamic adaptation" of Newroz by the Kurdish opposition forced the Turkish republic to try to "re-traditionalize" Nevruz, which was once almost forgotten in the official sphere. Although Nevruz, the Turkish version of Newroz, has historically been part of the traditional celebrations of Turkish culture, it was not marked as a major event on the Turkish national celebration calendar until the beginning of the 1990s.[35] A comprehensive list of the official and unofficial celebrations of Turkey in the year 1974 does not even list Nevruz (Akbayar 1975). The same situation can be seen in the Turkish Education Ministry's list of the celebrations that ought to be observed in the school system (*İlkokul, Ortaokul, Lise ve Dengi Okullar Eğitici Çalışmalar Yönetmeliği* 1983).

In contrast with these lists of national celebrations that made no mention of Nevruz, more recent compilations of social and cultural activities by the Prime Minister's Social and Cultural Affairs office (*Sosyal ve Kültürel Etkinlikler* 1996) and the Ministry of Culture's "Special Days Album of 1997" (Özel Günler Albümü 1997) included the Nevruz celebration. Thus, it was precisely at the point when a "bad" Kurdish Newroz began to have a significant political function that the Turkish state began to impose a new, "good" Turkish Nevruz.[36] The recent state-supported resurgence of "Nevruz consciousness" in Turkey is linked to the effort to create an official Turkish version of Newroz. This effort is aimed at co-opting the popularity and importance of Newroz as promulgated by the Kurds in Turkey. By recycling its traditional Nevruz and promoting its politicized, "good" version, the Turkish Ministry of Culture appropriated the tradition in an attempt to eclipse the popular Kurdish Newroz.

In this work I do not argue or search for the "pure" origin of Nevruz.[37] It is obvious that the tradition has deep roots and current applications within both Turkish and Kurdish populations in Turkey. Considering the tension surrounding this tradition however, it seems more pertinent to explore its contemporary utilization. Therefore, what I focus on here concerns the official appropriation of the festival, starting from 1991.

Two descriptions of the New Year celebration, both published anonymously, exemplify the official Turkish conceptualization of Nevruz. The first paragraph is taken from *Newspot* Magazine, which is published on behalf of the Directorate General of Press and Information and distributed worldwide. As part of state run public relations effort, *Newspot* covers wide range of topics, from foreign affairs to cultural issues and echoes the official stance:

> Nevruz is a feast celebrated in the Turkish world with great enthusiasm. It marks the rebirth of nature and renewal. Nevruz is a Persian word which means "New Day." The holiday occurs on March 21 . . . The beginning of spring was usually celebrated as a feast in the history of the Turks. Nevruz means the beginning of a new year, spring, mirth, love and friendship. It is the feast of peace, tolerance and friendship, symbolizing unity and togetherness. Nevruz originated in the Yenisey-Orhon area and spread by the Altaic Tribes. Then the Hun Turks brought this feast to Hungary and to the Balkans . . . Nowadays, it is celebrated by people who inhabit a vast area stretching from the Balkans to the Wall of China and to Siberia . . . In addition to the Huns, Kök Turks, Uigurs, Seljuks, and Ottomans, Nevruz has also been celebrated during the Republican period. Following the proclamation of the Republic, under Mustafa Kemal Atatürk, the holiday was celebrated as the Ergenekon Feast in 1922, 1923, 1924 and 1926 and later celebrations were held regionally. (*Newspot Magazine* 2002)[38]

Newspot's definition, written for the foreign general reader, situates Nevruz as a celebration with ancient Turkish roots that go back many centuries. The paragraph is triple coded with the deeply rooted connotations of official language. First, it prioritizes Turkic origins, and relegates Islam to a secondary role. Second, by presenting the Turks as the carriers of Nevruz tradition, the *Newspot* article provides a selective, ideologically refined map of the celebration. And finally, even though the article mentions the Persian roots of the word "Nevruz," it focuses mainly on Central Asian roots.

Employing a similar tone, the second passage, from a pamphlet produced by the Turkish Ministry of Culture, co-opts the tradition from a different angle. It stresses the traditional values of the Turks such as love for freedom, tolerance, friendship, and, most importantly, unity and togetherness. Underlining the solidarity among the Turkic nations and serving as a reminder of the historical difficulties that Turks faced, the quote suggests a specific set of meaning for the Nevruz tradition.

> Nawruz [occasionally used to refer to the official Nevruz], which is celebrated as the festival of liberation among Turks, is the Legend of Ergenekon related by Ebulghazi Bahadır in his work, *Geneological Tree of Turks*, an echo of the historical events quoted by old Chinese resources. It is the struggle for existence of Turks living for 400 years in a valley surrounded by high mountains. Returning to the fatherland from Ergenekon on a spring day, Turks won their liberty and independence and declared once more to friends and foes alike that they are still on their feet. (*Nawruz-New Day—March* 21 n.d.)

These texts represent the Turkish Ministry of Culture's response toward the tension-filled background of the tradition in Turkey. In short, they counteract the Kurdish claim by revitalizing and developing their own version, Nevruz. Equally important is the Turkish state's recognition that the multiple dimensions of this tradition could serve as a means of mitigating the current social and political situation. In other words, Nevruz not only gives the Turkish state an ideological outlet that could be used to reclaim a specific part of the cultural territory, but also creates a space for debating the key values of historical Turkish identity. During this reclamation process, the Turkish Ministry of Culture recycles its own version of the celebration from the traditional version, emphasizing togetherness and the importance of unity. This official version is presented as a pure Turkic tradition, with its unique context and connotations linking the early Asiatic Turkic origin with the Ottomans. This, in turn, sets the pace for a gravely needed national unity.

In the history of the Turkish republic, the promotion of Turkish culture not only marks the national identity but also excludes the existence of the other ethnicities living in Turkey. Through a variety of media, the state constantly supervised the "purity and safety" of the cultural sphere. However, this gate-keeping and its prompt application of dominance makes the cultural production process dependent on the political sphere.

Parallel to this heritage, starting from the 1990s, the intensive reinvention of Nevruz began in official circles. This timing also matches the heightening awareness of the popularity of Newroz among the Kurds in Turkey. In 1991, the General Directorate of Folk Culture Research and Development (HAGEM), one of the main service units of the Turkish Ministry of Culture, began to guide the rebirth of the Nevruz celebration all over Turkey. The Joint Administration of Turkic Culture and Arts (TŰRKSOY), an international foundation created for collaboration within the Turkish-speaking countries, also took part in some of these efforts. In a short period of time, the events surrounding Nevruz evolved into something on a more grandiose scale. It was then positioned and framed as an international festival of the Turkic people. This move was also related to the Turkish state's then popular coalition-making attempts with the newborn post-Soviet states of Central Asia. Participants and artists from all over Central Asia were invited to the celebrations.

The format of the celebrations starts taking shape months before the celebrations, when the Ministry of Culture sends form letters to the governors appointed by the national government, asking them to prepare a tentative program for the Nevruz celebrations in their cities. The local Nevruz Celebration Committee, usually led by the governor or the deputy governor, consists of managers and representatives from the local governmental agencies including the university. The Committee's task is to prepare a tentative schedule, which is then sent back to the Ministry of Culture for approval. These local programs generally are prepared according to local resources and priorities. Regional universities are considered crucial resources and take on significant roles in the events.

The Nevruz celebrations can be grouped into two major parts: indoors and outdoors. In contrast to Newroz, the government-sponsored Nevruz frames its festivities along more scholarly lines, assigning the public outdoor activities to a secondary position. Political leaders such as the Prime Minister and the President inaugurate the international conferences. Furthermore, on the basis of this ideological framework, state-supported TV and radio channels broadcast many programs on Nevruz and its celebrations. The Ministry of Culture lends support to the celebrations by sponsoring the publication of pictorial and coffee-table books. In 2001, for instance, the Ministry of Culture's Department of Education organized a painting, poetry, and essay contest for school children (*Çocuk Gözüyle Türk Kültüründe Nevruz* 2001).

From cartoon books to professional conferences, in the activities and state-supported publications, Nevruz is presented as a day of unity, love, and friendship. In all of these presentations Nevruz is cited as an original, purely Turkish tradition. By underscoring the Turkishness of the tradition, the state also organizes Nevruz as a bridge between the Turks of Turkey and the Turkic nations of the former Soviet Union. Although the Persian roots of the tradition are briefly mentioned in some of the literature, Nevruz is mainly positioned as having an Asiatic-Turkic origin. The Nevruz operation is also a pedagogical undertaking and reliant on the power of the various performances to affirm and eventually popularize its values. It is used not only to express anti-Newroz messages but also to promote and enrich the Turkish national identity. Thus, it becomes part of a coherent effort to control the representation of the past and is directly related to the pedagogical affairs.

What happens in these activities is, to some extent, a manifestation of a pedagogical project to construct the Turkish identity as a unifying framework. In this process, the Turkish Ministry of Culture's agencies and the related independent organizations redefine and use the symbols of Nevruz to provide a counter venue to link the concept of Nevruz to Turkish identity.

Even though the Turkish identity dominates Nevruz celebrations and associated activities, it is not the main focus of those celebrations. It is crucial to differentiate between the two major functions of the Nevruz-centered pedagogy. The first role, as I mentioned above, is the presentation of the Turkish identity as a framework for the multiethnic society of Turkey. Although this comes into view as a main theme, the unlearning of Newroz is also a key component of the process. Since the official stance considers Newroz as an outcome of divisive activities of and promoted by terrorist organizations that contradicts and challenges the existence of the Turkish republic, it hopes to fix this false conscioussness through the promotion of Nevruz, the official "good" Turkish version.

The high level of politicization of the New Year celebrations in Turkey makes the event a significant one. By drawing on shared symbols and pieces of history, both Newroz and Nevruz "establish continuity with a suitable historic past." As Eric Hobsbawm and Terence Ranger observed, traditions do not pass completely unchanged through the generations, but are often interpreted and reinterpreted to satisfy the needs of the current generation (Hobsbawm and Ranger 1983, p. 1). In instances like Nevruz, the state remakes and attempts to reclaim a tradition, especially following its success on the popular level. The pedagogy that emerges out of this operation is what Freire would call "the pedagogy of domestication." It is a pedagogy that continuously repeats itself in the unlearning of the Kurdish Newroz. When symbols such as the Nevruz are institutionalized and reiterated through the "state ideological apparati" (Althusser 1971), such as educational systems and religious institutions, the officially endorsed symbols and meanings become incorporated as part of everyday lived experience. This goes a long way in "maintaining the 'official' universe against the heretical challenge" (Berger and Luckmann 1967, p. 99) and in the socialization of the population to accept the official Nevruz version as the only "true" one. Nevruz and its associated activities, then, are "learned as objective truth in the course of socialization and thus internalized as subjective reality" (Berger and Luckmann 1967, p. 62).

It is in this way that the Turkish pedagogy sets itself in direct opposition to the Kurdish identity as presented by Kurdish political organizations in the past and present. The Nevruz, thus, stands as an example of the idea that educational practice and its theory cannot be neutral. "The relationship between practice and theory in education oriented toward liberation is one thing, but quite another in education for the purpose of domestication" (Freire 1970, p. 12). Nevruz provides the Turkish state with the opportunity and flexibility to popularize and circulate its own message. On this basic level, the Nevruz pedagogy is not only designed to shape the consciousness of the audience, but also to establish itself as the "good" domesticating message.

CONCLUSION

This chapter examines two distinct pedagogical recycling processes of New Year celebrations in Turkey. The historical background, which fostered both the unofficial Kurdish civic education and the official Turkish pedagogy, has a crucial position in these pedagogical practices. The key factors fueling contemporary conflict in the cultural sphere may be traced to unresolved problems from the Turkish modernization/Westernization project and the ingrained "continuity" within the Turkish cultural revolution. I also argue that this conflict arose from mutual pedagogical borrowings. While, on the one hand, Kurdish nationalists learned from Kemal Atatürk's continuous revolution to construct an alternative civic education, on the other hand, Turkish authorities also came to realize the utility of Nevruz as a pedagogical and nation-building tool through the popular success of the Kurdish Newroz.

This study suggests that starting from the late 1970s, the Kurdish opposition employed a traditional narrative to define and justify the Kurds' historical rights. By making the Newroz tradition one of its crucial pedagogical bases, the Kurdish leaders essentially created socialization channels to aid the dissemination of the Kurdish identity. Although, on the surface, Newroz may appear to have a narrow political operation, its enduring aspects, such as the leading role of the youngsters in the collective education process and the promotion of the "new" text, challenge the traditional learning/teaching processes.

With the Turkish reclamation of Newroz that resulted in the rebirth of Nevruz, a new venue has come into being to compete with the success of the Kurdish Newroz and to secure the current stance of the Turkish state in southeastern Turkey. Besides making its presence felt on the cultural front, the Nevruz framework is also intended to bridge the Asiatic Turkic people of the former Soviet republics with the Turkish republic for international, political purposes. The promotion of "unlearning" of the opposite projects and, in Zygmunt Bauman's words, preventing the identities from sticking became the common characteristics in both cases (Bauman 1996, p. 24). While the Kurdish Newroz was born to highlight and distance the Kurdish identity from the dominant Turkish identity, the Nevruz was also intended to do the same, but in reverse.

This chapter has established a new understanding of the Newroz and Nevruz celebrations. I believe there is a significantly positive outcome from such an understanding. Despite the opposing interpretations of the New Year festival, these celebrations provide a common cultural ground on which negotiations between the two sides may take place to promote a peaceful future. As one of the very limited number of venues that brings "the Kurdish question" into the spheres of culture and education in Turkey, Newroz/Nevruz has a good potential of becoming a major venue for dialogue. The festivals can also give an opportunity to the various actors to examine prejudgments for the "other" side.

On another plane, being linked to the "Kurdish question," one of the key tensions in the Middle East, Nevruz/Newroz also provides insight into opportunities inherent in the revolution/pedagogy junction from outside the school. However irregular and problematic the Newroz/Nevruz presentations may be, they demonstrate the extent to which pedagogical operations may be used in a productive manner to promote a model for long-term reconciliation. It indicates that if members on opposite sides can construct pedagogies that form a bridge across communities, both civic education at the official level and informal teaching within communities may be used to promote a long-term reconciliation between the two camps.

Remember the imaginary visitor of one of the major cities of Turkey on any given March 21 at the opening of my chapter. As the findings of this research reveal, the visitor could come to observe an integrated celebration on some March 21 in the near future, where the people will recognize and respect their differences.

Notes

The fieldwork of this research was assisted by the Mershon Center Grant (Summer 2001 and Spring 2002) and The Ohio State University, Graduate Student International Dissertation Research Travel Grant (Summer 2000). I am indebted to all who supported this research in various forms in the field. I should also thank Nicole Duran, Dorothy Noyes, Selina Lim, and the volume's editor Tom Ewing, who gave me valuable suggestions for strengthening the paper's thesis and improving its structure.

1. The rendition of the festival's name has been very controversial in Turkey. "Nevruz" stands for the Turkish version. "Newroz" represents the Kurdish version. However, since the Turkish alphabet does not include the letter "w," the use of Newroz is prohibited. To protect themselves from "w" related persecution, some writers use "v" instead of "w" and write Newroz as, Nevroz. For an example of a "w" related court case, see Alataş (2003, p. 90).
2. See Aksoy (1998) and Çay (1999) for their most comprehensive historical accounts.
3. This way of conceptualizing identity creation processes, I hope, does not give an impression that I disregard the complexity and the importance of other dynamics in play.
4. I am familiar with the risks of comparing the whole pedagogical practices of the state through Nevruz and the outcome of Newroz. However, the richness of the similarities between the two and the originality of the political competition convince me to further the argument in this direction. Consistent with this preference and to recognize the differences between the two practices I use "civic education" to describe the outcome of the Kurdish Newroz to show its lesser position.
5. While inspecting the Newroz related learning practices that was first touched upon by Yılmaz Varol (1994), I encountered with David Tyack's argument on the private shaping of civic educations (2001, p. 333), Lawrence Blum's conceptualization of "antiracist civic education" (1996, p. 23), and James Leith's eloquent presentation on the role of the printed images in civic education during the French Revolution

(1996). These all helped me to refine my thoughts on the nonstate and nontraditional actors of the pedagogical processes.

6. For a good account of the connection between the recycling and the cultural memory, see Neville and Villeneuve (2002).

7. The peninsula of Anatolia (*Anadolu*) is considered the heartland of Turkey. For brief information on the region, see Howard (2001, pp. 6–7).

8. The last census to include the question of language, which functioned to show native speakers of other languages including Kurdish, was conducted in 1965. Because of the official census policies of Turkey, there is no accurate demographic data available to know the Kurdish population of Turkey exactly. For the history of the census in Turkey, see Dündar (1999). For an example of an assessment which was produced on the available data see Andrews (1989) and (2002), Özsoy et al. (1992), Mutlu (1995, 1996). See also Frey (1965, p. 147, footnote 23) for an early example of lack of statistical data on Kurds in Turkey [Frey's note is taken from Jafar 1976, p. 44].

9. For a general discussion on the effects of Ottoman past and Islamic heritage on post-Ottoman identity creation in Turkey, see Findley (2000).

10. According to Mesut Yeğen, the exclusion of the Kurdish identity was a result of the Turkish nation-creation project (Yédev 1996). For a brief but comprehensive discussion of the Kurds' exclusion, see Güneş-Ayata and Ayata (1999, pp. 130–131).

11. In the Kurdish terminology, *agha* defines the leaders of a tribe or clan. It implies landlordship and is also used for the children and brothers of the actual landlord. In current use, it defines the one who rules (Bruinessen 1992, p. 80; Kinnane 1964, p. 11). For an informative argument of Kurdish tribalism and its evolution, see White (2000, pp. 17–22).

12. For a comprehensive account of the tension between the various factions of the Kurdish nationalism in that period, see Özoğlu (2004).

13. For an insightful analysis of the construction of the Turkish culture, see Navaro-Yashin (2002).

14. This point does not disregard the deep historical roots of Newroz within the Kurdish culture. For instance, Ely Bannister Soane in his account of the journey that he made between 1907–1909 mentions a legend on the origin of the Kurds which resembles the current Newroz framework (Soane 1979 [1912], p. 368). Nicole Watts notes how the leaders of Dersim uprising of 1937 chose Newroz to start their attacks (Watts 2000, p. 21). Thomas Bois also remarks the celebration of Newroz by the Kurds in much earlier periods (Bois 1966, pp. 7, 67). For an analysis of the position of Newroz in the Kurdish culture before the 1970s, see Aksoy (1998) and Cengiz (2003).

15. Obviously, the course of popularity within the Kurdish population is much more complex than this sentence states but has to be covered in another work.

16. Babcock (1978), Cantwell (1992, 1993), Fernandez (1986b), Geertz (1971, 1993 [1973]), Turner (1982, 1986), Freire (1970, 1985), and Gore (1993) are central to my analysis.

17. The Kemalism is the tenets that the Turkish republic founded upon. For a brief definition, see Howard (2001, pp. 107–108). For an analysis of its contemporary meaning, see Mango (2002a).

18. Dietrich Jung argues the success of this campaign that "most social structures, especially in the periphery" was not affected by these reforms (2002, pp. 78–79). By claiming the lack of mass support and Kemalist dependence upon the traditional leaders as brokers between center and periphery, he finds the outcome

of these reforms questionable. For a similar comment, see Poulton (1997, pp. 121–122).

19. See, e.g., Öztürkmen (1994).

20. According to Navaro-Yashin reminiscence of a Turkic past in Central Asia was not existent among people of Anatolia in the early twentieth century (2002, p. 11).

21. In his 1998 article "From Ottoman to Turk: Self-Image and Social Engineering in Turkey" historian Selim Deringil criticizes this view and considers the comments on the inferior position of Turks during the Ottoman era as part of "the standard jargon of Turkish official historiography" (Deringil 1998, p. 217). Against the works of Ahmad that I mentioned above, Berkes (1964) and Lewis (2002 [1961]), Deringil brings the findings of more recent research showing that "even in the 'Golden Age' of the empire," the ruling elite had a clear notion of their Turkic identity and were proud of it (Fleisher 1986, pp. 286–290). Since my research is more about the implementation and the results of the great Turkification project of the republican era, I disregard Deringil's valuable challenge at least for this work.

22. For the minutes of the meetings see *Birinci Türk Tarih Kongresi, Konferanslar, Müzakere Zabıtları* 1932. For a comprehensive account of the Congress, see Behar (1992).

23. For details of the area, which was designated as Kurdistan, and the articles of 62, 63, and 64 of the Treaty of Sevres relating to it, see McDowall (2000, pp. 464–465), Kurubaş (2004, pp. 99–100), and Bayrak (1993, pp. 101–108).

24. For an outline of the diplomatic history of the period, see Macfie (1996). For a detailed analysis of the position of minorities in the Treaty of Lausanne, see Ürer (2003).

25. For instance, in the 1924 constitution, citizenship called for an explicit acceptance of Turkishness (Barkey and Fuller 1998, p. 10).

26. There is a continuous debate on the nature of the Kurdish rebellions. While some consider these as religious uprisings started by reactionaries, others explain the religious nature of the early Kurdish movements by the priority of wide coalition making. See Olson (1989, p. 153), Mumcu (1991), Ciment (1996, p. 46), and Bozarslan (1992, p. 96).

27. Though I agree with Hugh Poulton's point on the crucial position of the Marxist elite for given period, and the left's failure in the 1965 and 1969 elections in Turkey, some concurrent developments require a reconsideration of his argument for how they "failed to attract widespread support and failed miserably" (Poulton 1997, pp. 210–211). For instance, massive "Eastern Meetings" which took place in the autumn of 1967 in various southeastern, dominantly Kurdish cities, do not only show the growing popular support for the Kurdish opposition, but also help us to see the strong internal dynamics within these organizations (Beşikçi 1992, p. 64; White 2000, p. 133). The Workers' Party of Turkey became the first legal party to recognize and bring the Kurdish question to the national political arena, which was a groundbreaking shift in the social and political climate for further pro-Kurdish actions (Barkey and Fuller 1998, p. 15; Poulton 1997, pp. 211–212).

28. This situation continued up to the imprisonment of PKK President Abdullah Öcalan in February 1999. Consistent with the consequent decline of the armed confrontation after his arrest, Newroz celebrations begun to be more conciliatory.

29. According to the Internet editions of daily *Milliyet*, November 11, 2003 (last minute section) and *Özgür Politika*, November 12, 2003, with its eighth congress, the PKK changed its name and became "Kongra Azadî û Demokrasiya Kürdistan,

KADEK" (Kurdistan Freedom and Democracy Congress), in April 4, 2002. [Daily Milliyet's Internet edition in November 12, 2003 gives April 16, 2002, for the date for the change. Gunter (2004, p. 121) provides February 2002 for this change.] KADEK in its second congress in October 26, 2003 revocated itself. After this move the movement took the name of "Kongra Gelê Kurdistan, KONGRA-GEL" (Kurdistan People's Congress) in October 27, 2003. Considering these constant shifts, I preferred to stick with the use of PKK. Although Mustafa Karasu, one of the members of the presidential council of PKK, explains these changes as breaking away from the hierarchical centralist models and steps toward reorganization and modernization of the party (Karasu 2004), European Union and the United States' decision to put PKK (and the new versions) into the lists of terrorist organizations must be considered in these proceedings.

30. For details of the conditions in Diyarbakır Prison No. 5, see Zana (1997), Ural (1988), and Cemal (2003).
31. *Yeniden Özgür Gündem*, March 22, 2003.
32. The "national ownership" of this color combination, red-green-yellow, is another controversial and contested issue. See Tural and Kılıç (1996).
33. For specifically previous Newroz related human rights violations see *İHD Şube ve Temsilciliklerinin Olağanüstü Hal Bölge Raporu* 1992, pp. 9–77.
34. I first encountered this point in Aksoy (1998, pp. 179–180).
35. By making this point I do not claim that Nevruz has not been part of both traditional and official celebrations in Turkey. What I claim is that although Nevruz was celebrated officially by the young republic for a short period of time, it eventually faded away until the success of the Kurdish Newroz in Turkey.
36. I borrowed these words from journalist Kevin McKiernan through his 2000 documentary: *Good Kurds, Bad Kurds*.
37. Further information on Nevruz can be seen in Çay (1999), Pekcan and Öztürk (1993), Başbuğ (1985), Makas (1987), and Eker and Abatoğlu (1993).
38. The same text can be also found in the website of the Turkish Ministry of Culture at <www.kultur.gov.tr>. For similar texts under the title "Nevruz: The World's Oldest Festival," see *Newspot Magazine* (2000, pp. 36–37).

References

Ahmad, Feroz. 1993. *The Making of Modern Turkey*. New York: Routledge.

———. 2003. *Turkey: The Quest for Identity*. Oxford: Oneworld Publications.

Akbayar, Yaman. 1975. *1 Yılın Kılavuzu 1974*. İstanbul: Fakülteler Matbaası.

Aksoy, Gürdal. 1998. *Bir Söylence Bir Tarih: Newroz*. Ankara: Yurt Kitap-Yayın.

Alataş, Evrim. 2003. "Nevroz, Nevruz ve Newroz." In *Mayoz Bölünme Hikayeleri*. Istanbul: Aram Yayıncılık, pp. 85–90.

Althusser, Louis. 1971. "Ideology and Ideological State Apparatuses." in *Lenin and Philosophy and Other Essays*. Translated by Ben Brewster. New York: Monthly Review Press.

Anderson, Benedict. 1991. *Imagined Communities: Reflections on the Origin and Spread of Nationalism*. London: Verso.

Anderson, Charlotte C. and James H. Landman. 2003. *Globalization and Border Crossings: Examining Issues of National Identity, Citizenship, and Civic Education*. Chicago: American Bar Association.

Andrews, Peter Alford, ed. 1989/2002. *Ethnic Groups in the Republic of Turkey*. Wiesaden: Dr. Ludwig Reichert Verlag.

Attar, Ali. 1998. "Nawruz in Tajikistan: Ritual or Politics?" In *Post Soviet Central Asia*. Touraj Atabaki and John O'Kane, eds. London: Tauris & Co Ltd.

Aydın, Zülküf. 2002. "Uncompromising Nationalism: The Kurdish Question in Turkey." In *The Politics of Permanent Crisis: Class, Ideology and State in Turkey*. Neşecan Balkan and Sungur Savran, eds. New York: Nova Science Publishers, pp. 85–105.

Babcock, Barbara A. 1978. *The Reversible World: Symbolic Inversion in Art and Society*. Ithaca: Cornell University Press.

Barkey, Henri J. and Graham E. Fuller. 1998. *Turkey's Kurdish Question*. New York: Rowman & Littlefield.

Başbuğ, Hayri. 1985. "Nevruz." *Türk Dünyası Araştırmaları*. Şubat.

Bauman, Zygmunt. 1996. "From Pilgrim to Tourist or a Short History of Identity." In *Questions of Cultural Identity*. Stuart Hall and Paul du Gay, eds. London: Sage Publications, pp. 18–36.

Bayrak, Mehmet. 1993. *Kürtler ve Ulusal-Demokratik Mücadeleleri*. Ankara: Özge Yayınları.

Behar, Büşra Ersanlı. 1992. *İktidar ve Tarih 1929–1932*. İstanbul: Afa Yayınları.

Berger, Peter L. and Thomas Luckmann. 1967. *The Social Construction of Reality: A Treatise in the Sociology of Knowledge*. Garden City: Anchor Books.

Berkes, Niyazi. 1964. *The Development of Secularism in Turkey*. Montreal: McGill University Press.

Beşikçi, İsmail 1992. *Doğu Mitingleri'nin Analizi (1967)*. Ankara: Yurt Kitap-Yayın.

Birinci Türk Tarih Kongresi, Konferanslar, Müzakere Zabıtları 1932. İstanbul: Matbabacılık ve Neşriyat Türk Anonim Şirketi.

Blum, Lawrence A. 1996. "Antiracist Civic Education in the California History-Social Science Framework." In *Public Education in a Multicultural Society: Policy, Theory, Critique*, Robert K. Fullinwider, ed. New York: Cambridge University Press, pp. 23–48.

Bois, Thomas. 1966. *The Kurds*. Translated from French by M. W. M. Welland. Beirut: Khayats.

Bozarslan, Hamit. 1992. "Political Aspects of the Kurdish Problem in Contemporary Turkey." In *The Kurds: A Contemporary Overview*. Philip G. Kreyenbroek and Stefan Sperl, eds. New York: Routledge, pp. 95–114.

Bruinessen, Martin van. 1990. "Kurdish Society, and Modern State: Ethnicity, Nationalism Versus Nation-Building." In *Kurdistan in Search of Ethnic Identity*. Touraj Atabaki and Margreet Dorleijn, eds. Utrecht: Houtsma Foundation Publication Series, pp. 24–51.

———. 1992. *Agha, Sheikh and State: The Social and Political Structures of Kurdistan*. London and New Jersey: Zed Books.

———. 1994. "Genocide in Kurdistan: The Suppression of the Dersim Rebellion in Turkey (1937–38) and the Chemical War Against the Iraqi Kurds (1988)." In *Genocide: Conceptual and Historical Dimensions*. George J. Andreopoulos, ed. Philadelphia: University of Pennsylvania Press.

———. 2000. "The Kurds in Turkey." In *Kurdish Ethno-Nationalism Versus Nation-Building State (Collected Articles)*. Istanbul: The ISIS Press.

Cantwell, Robert. 1992. "Feasts of Un-naming: Folk Festivals and the Representation of Folklife." In *Public Lore*. Robert Baron and Nicholas Spitzer, eds. Washington, D.C.: Smithsonian.

———. 1993. *Ethnomimesis: Folklife and the Representation of Culture*. Chapel Hill: University of North Carolina Press.

Cemal, Hasan. 2003. *Kürtler.* İstanbul: Doğan Kitap.

Cengiz, Daimi. 2003. "Kürt Şiiri ve Müziğinde Newroz." *Evrensel Kültür* Mayıs. Sayı 137, pp. 50–53.

Chailand, Gerard, ed. 1993. *A People Without a Country: The Kurds and Kurdistan.* Translated by Michael Pallis. New York: Olive Branch Press.

———. 1994. *The Kurdish Tragedy.* Translated by Philip Black. Atlantic Highlands: Zed Books.

Ciment, James. 1996. *The Kurds: State and Minority in Turkey, Iraq and Iran.* New York: Facts on File.

Çay, Abdulhaluk M. 1999. *Nevruz: Türk Ergenekon Bayramı.* Ankara: Tamga Yayıncılık.

Çocuk Gözüyle Türk Kültüründe Nevruz 2001. Ankara: Kültür Bakanlığı Eğitim Dairesi Başkanlığı.

Davison, Roderic H. 1998. *Turkey: A Short History.* Third edition updated by Clement H. Dodd. Huntingdon, England: The Eothen Press.

De Vos, George A. and Lola Romanucci-Ross. 1995. "Ethnic Identity: A Psychocultural Perspective." In *Ethnic Identity: Creation, Conflict, and Accommodation.* Lola Romanucci-Ross and George A. De Vos, eds. London: Altamira Press, pp. 349–379.

Deringil, Selim. 1998. "From Ottoman to Turk: Self Image and Social Engineering in Turkey." In *Making Majorities: Constituting the Nation in Japan, Korea, China, Malaysia, Fiji, Turkey, and the United States.* Dru C. Gladney, ed. Stanford: Stanford University Press, pp. 217–226.

Dündar, Fuat. 1999. *Türkiye Nüfus Sayımlarında Azınlıklar.* İstanbul: Doz Yayınları.

Eker, Süer and Ahmet Abatoğlu. 1993. *Nevruz: Ulusun Ulu Günü.* Ankara: Ajans L.

Fernandez, James W. 1986a. "Folklorists as Agents of Nationalism: Asturian Legends and the Problem of Identity." In *Fairy Tales and Society: Illusion, Allusion, and Paradigm.* Ruth B. Bottigheimer, ed. Philadelphia: University of Pennsylvania Press.

Fernandez, James W. 1986b. *Persuasions and Performances: The Play of Tropes in Culture.* Bloomington: Indiana University Press.

Fırat, Ümit. 2003. "Hasan Cemal in Kürtleri." *Virgül.* June 3.

Findley, Carter V. 2000. "Continuity, Innovation, Synthesis, and the State." In *Ottoman Past and Today's Turkey*, Kemal Karpat, ed. Leiden, The Netherlands: Brill.

Fleisher, Cornell H. 1986. *Bureaucrat and Intellectual in the Ottoman Empire: The Historian Mustafa Ali (1541–1600).* Princeton, N.J.: Princeton University Press.

Freire, Paulo. 1970. *Cultural Action for Freedom.* Monograph Series No. 1. Cambridge: Harvard Educational Review.

———. 1985. *The Politics of Education: Culture, Power, and Liberation.* New York: Bergin and Garvey.

Frey, Frederick W. 1965. *The Turkish Political Elite.* Cambridge: MIT Press.

Fuller, Graham E. 1999. "Turkey's Restive Kurds: The Challenge of Multiethnicity." In *Ethnic Conflict and International Politics in the Middle East.* Leonard Binder, ed. Gainesville: University of Florida Press.

Geertz, Clifford. 1971. *Myth, Symbol and Culture.* New York: Norton.

———. 1984. "From the Native's Point of View: On the Nature of Anthropological Understanding." In *Culture Theory: Essays on Mind, Self, and Emotion.* Richard A. Shweder and Robert A. LeVine, eds. New York: Cambridge University Press.

———. 2000. "Deep Play: A Description of the Balinese Cockfight." In *The Interpretation of Culture.* New York: Basic Books.

Geldern, James von. 1993. *Bolshevik Festivals, 1917–1920.* Berkeley; London: University of California Press.

Gellner, Ernest. 1983. *Nation and Nationalism*. Ithaca, N.Y.: Cornell University Press.

————. 1994. "Kemalism." In *Encounters with Nationalism*. Cambridge: Blackwell.

Glasse, Cyril. 2001. *The New Encyclopedia of Islam*. New York: Rowman and Littlefield.

Goffman, Erwin. 1961. *Encounters: Two Studies in the Sociology of Interaction*. Indianapolis: Bobbs-Merril.

Gore, Jennifer. 1993. *The Struggle for Pedagogies: Critical and Feminist Discourses as Regimes of Truth*. New York: Routledge.

Gunter, Michael M. 2004. *Historical Dictionary of the Kurds*. Lanham, MD: Scarecrow Press.

Gülalp, Haldun. 1997. "Modernization Policies and Islamist Politics in Turkey." In *Rethinking Modernity and National Identity in Turkey*. Sibel Bozdoğan and Reşat Kasaba, eds. Seattle: University of Washington Press.

Güneş-Ayata, Ayşe and Sencer Ayata. 1999. "Ethnicity and Security Problems in Turkey." In *New Frontiers in Middle East Security*. Lenore G. Martin, ed. pp. 127–150.

Hall, Stuart. 1996. "Introduction: Who Needs 'Identity'?" In *Questions of Cultural Identity*.

Henderson, Helene and Sue Ellen Thompson. 1997. *Holidays, Festivals and Celebrations of the World Dictionary*. Detroit: Omnigraphics.

Hobsbawm, Eric. 1992. *The Nations and Nationalism since 1780: Programme, Myth, Reality*. Cambridge: Cambridge University Press.

———— and Terence Ranger, eds. 1983. *The Invention of Tradition*. Cambridge: Cambridge University Press.

Houston, Christopher. 2001. *Islam, Kurds and the Turkish Nation State*. Oxford; New York: Berg.

Howard, Douglas A. 2001. *The History of Turkey*. Westport, Conn.; London: Greenwood Press.

İHD Şube ve Temsilciliklerinin Olağanüstü Hal Bölge Raporu. 1992. Diyarbakır: İnsan Hakları Derneği Diyarbakır Şubesi Yayınları.

İlkokul, Ortaokul, Lise ve Dengi Okullar Eğitici Çalışmalar Yönetmeliği. 1983. Ankara: Milli Eğitim Basımevi.

Jafar, Majeed R. 1976. *Under-Development: A Regional Case Study of the Kurdish Area in Turkey*. Helsinki, Finland: Painoprint Oy.

Jung, Dietrich with Wolfango Piccoli. 2001. *Turkey at the Crossroads, Ottoman Legacies and A Greater Middle East*. London; New York: Zed Books.

Kabacalı, Alpay. 1991. *Tarihimizde Kürtler ve Ayaklanmaları*. İstanbul: Cem Yayınevi.

Karasu, Mustafa. 2004. "Kongra-Gel'in Yaşam ve Mücadele Rehberi Önderliktir." *Özgür Halk*. 15 Ocak. Yıl 14. Sayı 146, pp. 57–60.

Kili, Suna. 2003. *The Atatürk Revolution: A Paradigm of Modernization*. İstanbul: Türkiye İş Bankasi Kültür Yayınları.

Kinnane, Derk. 1964. *The Kurds and Kurdistan*. London; New York: Oxford University Press.

Kinross, Patrick Balfour. 1965. *Atatürk: A Biography of Mustafa Kemal, Father of Modern Turkey*. New York: William Morrow.

Kirişçi, Kemal and Gareth M. Winrow. 1997. *The Kurdish Question and Turkey: An Example of a Trans-State Ethnic Conflict*. London: Frank Cass.

Kurubaş, Erol. 2004. *Sevr-Lozan Sürecinden 1950'lere Kürt Sorununun Uluslararaş Boyutu ve Türkiye* (2 vols.). Ankara: Nobel.

Lebor, Adam. 1997. *A Heart Turned East: Among the Muslims of Europe and America*. London: Little, Brown and Company.

Leith, James. 1996. "Ephemera: Civic Education Through Images." In *The French Revolution in Social and Political Perspective*. Peter Jones, ed. New York: Arnold, pp. 188–202.

Lewis, Bernard. 1968 [1961]. *The Emergence of Modern Turkey*. Oxford: Oxford University Press.

Lipovsky, Igor P. 1992. *The Socialist Movement in Turkey 1960–1980*. Leiden: E. J. Brill.

MacDonald, Margaret Read, ed. 1992. *The Folklore of World Holidays*. Detroit: Gale Research Inc.

Macfie, A. L. 1996. *The Eastern Question, 1774–1923*. London; New York: Longman.

Makas, Zeynelabidin. 1987. "Sözlü ve Yazılı Kaynaklarda Nevruz." In *Türk Milli Kültüründe Nevruz*. İstanbul: Türk Dünyası Araştırmaları Vakfı Yayını.

Mango, Andrew. 2000. *Atatürk*. New York: The Overlook Press.

———. 2002a. "Kemalism in a New Century." In *Turkish Transformation New Century—New Challenges*. Brian Beeley, ed. Cambridgeshire: Eothen Press, pp. 22–36.

———. 2002b. "Atatürk: Founding Father, Realist, and Visionary." In *Political Leaders and Democracy in Turkey*. Metin Heper and Sabri Sayarı, eds. Oxford: Lexington Books, pp. 9–24.

Mardin, Şerif. 1981. "Religion and Secularism in Turkey." In *Atatürk: Founder of a Modern State*. Ali Kazancıgil and Ergun Özbudun, eds. London: C. Hurst.

McDowall, David. 1992. *The Kurds: A Nation Denied*. London: Minority Rights Publications.

———. 2000. *A Modern History of the Kurds*. London: Tauris.

Mead, Margaret. 1978. *Culture and Commitment: The New Relations Between the Generations in the 1970s* (second edition). New York: Columbia University Press.

Mumcu, Uğur. 1991. *Kürt Islam Ayaklanması*. Ankara: Tekin Yayınevi.

Murrell Jr., Peter C. 2002. *African Centered Pedagogy: Developing Schools of Achievement for African American Children*. New York: SUNY Press.

Mutlu, Servet. 1995. "Population of Turkey by Ethnic Groups and Provinces." *New Perspectives on Turkey* vol. 12, pp. 33–60.

———. 1996. "Ethnic Kurds in Turkey: A Demographic Study." *International Journal of Middle East Studies* vol. 28, no. 4, pp. 517–541.

Myerhoff, Barbara. 1982. "Rites of Passage: Process and Paradox." In *Celebration: Studies in Festivity and Ritual*. Victor Turner, ed. Washington, D.C.: Smithsonian Institution Press, pp. 109–135.

Nachmani, Amikam. 2003. *Turkey: Facing a New Millennium, Coping with Intertwined Conflicts*. Manchester: Manchester University Press.

Navaro-Yashin, Yael. 2002. *Faces of the State: Secularism and Public Life in Turkey*. Princeton: Princeton University Press.

Nawruz-New Day—March 21. N.D. Published by the Ministry of Culture, General Directorate of Research and Development of Folk Cultures.

Neville, Brian and Johanne Villeneuve, eds. 2002. *Waste-Site Stories: The Recycling of Memory*. Albany: State University of New York Press.

Newspot Magazine. 2000. "Nevruz: The World's Oldest Festival," no. 20 (March–April), pp. 36–37.

Newspot Magazine. 2002. "Nevruz," no. 32 (March–April), p. 43.

O'Ballance, Edgar. 1973. *The Kurdish Revolt 1961–70*. London: Faber and Faber.

Olson, Robert. 1989. *The Emergence of Kurdish Nationalism and the Sheikh Said Rebellion, 1880–1925*. Austin: University of Texas Press.

Oran, Baskın. 1996. "Kürt Milliyetçiligi: Doğuşu ve Gelişmesi." In *Türkiye'nin Kürt Sorunu.* Seyfettin Gürsel, ed. İstanbul: Türkiye Sosyal Araştırmalar Vakfı.

Owen, Roger. 1997. "Modernizing Projects in Middle Eastern Perspective." In *Rethinking Modernity and National Identity in Turkey.* Sibel Bozdoğan and Reşat Kasaba eds. Seattle: University of Washington Press.

Özel Günler Albümü. 1997. Ankara: Kültür Bakanlığı Araştırma Planlama ve Koordinasyon Kurulu Başkanlığı.

Özoğlu, Hakan. 2004. *Kurdish Notables and the Ottoman State.* New York: State University of New York Press.

Özsoy, A. E., I. Koç, and A. Toros. 1992. "Türkiye'nin Etnik Yapısının Ana Dil Sorularına Göre Analizi." *Nüfusbilim Dergisi,* 14.

Öztürkmen, Arzu. 1994. "The Role of People's Houses in the Making of National Culture in Turkey." *New Perspectives on Turkey* (Fall), pp. 159–181.

Pekcan, Yıldız and Sevinç Öztürk. 1993. *Tarih ve Etnoğrafya Açısından Nevruz.* Ankara: Boğaziçi Yayınları.

Perinçek, Doğu. 1999. *Kurtuluş Savaşında Kürt Politikası.* İstanbul: Kaynak Yayınları.

Polat, Edip. 1991. *Newrozladık Şafakları.* Ankara: Başak Yayınları.

Poulton, Hugh. 1997. *Top Hat, Grey Wolf and Crescent: Turkish Nationalism and the Turkish Republic.* New York: New York University Press.

Shariati, Ali. 1986. "Nowruz." *Iranian Studies* vol. 19, nos. 3–4, pp. 235–241.

Shaw, Stanford J. and Ezel Kural Shaw. 1977. *History of the Ottoman Empire and Modern Turkey.* Cambridge: Cambridge University Press.

Soane, Ely Bannister. 1979 [1912]. *To Mesopotamia and Kurdistan in Disguise: Narrative of a Journey from Constantinople through Kurdistan to Baghdad, 1907–1909.* Amsterdam: Academic Publishers Associated.

Sosyal ve Kültürel Etkinlikler. 1996. Ankara: Başbakanlık Sosyal ve Kültürel İşler Başkanlığı.

Thompson, Sue Ellen, ed. 1998. *Holiday Symbols.* Detroit: Omnigraphics.

Torney-Purta, Judith, John Schwille and Jo-Ann Amadeo, eds. 1999. *Civic Education Across Countries: Twenty-four National Case Studies from the IEA Civic Education Project.* Delft: Eburon Publishers.

Tural, Sadık and Elmas Kılıç, eds. 1996. *Nevruz ve Renkler: Türk Dünyasında Nevruz İkinci Bilgi Şöleni Bildirileri.* Ankara: Atatürk Kültür Merkezi Yayını.

Turner, Victor. 1982. *Celebrations: Studies in Festivity and Ritual.* Washington, D.C.: Smithsonian Institution Press.

———. 1986. *The Anthropology of Performance.* New York: PAJ Publications.

Tyack, David. 2001. "School for Citizens: The Politics of Civic Education from 1790 to 1990." In *E Pluribus Unum? Contemporary and Historical Perspectives on Immigrant Political Incorporation.* Gary Gerstle and John Mollenkopf, eds. New York: Russel Sage Foundation, pp. 331–370.

Ural, M. Ali. 1988. *Sorgu, Zindan, Direniş ve Yasam.* Diyarbakır: Dilan Yayinlari.

Ürer, Levent. 2003. *Azınlıklar ve Lozan Tartışmaları.* İstanbul: Derin Yayınları.

Varol, Yılmaz. 1994. "Kawanın Kimliği Üzerine Birkaç Not." *Yeni İnsan.* Kasım.

Watts, Nicole. 2000. "Relocating Dersim: Turkish State-Building and Kurdish Resistance, 1931–1938." *New Perspectives on Turkey.* (Fall), pp. 5–30.

White, Paul J. 2000. *Primitive Rebels or Revolutionary Modernizers? The Kurdish National Movement in Turkey.* London; New York: Zed Books.

Yeğen, Mesut. 1996. "The Turkish State Discourse and the Exclusion of Kurdish Identity." In *Turkey: Identity, Democracy, Politics.* Sylvia Kedourie, ed. Portland: Frank Cass.

Yeniden Özgür Gündem. March 22, 2003.

Zana, Mehdi. 1991. *Bekle Beni Diyarbakır.* Hazırlayan Ali Öztürk. İstanbul: Doz Yayıncılık.

———. 1997. *Prison No. 5: Eleven Years in Turkish Jails/As Told to Andre Vauquelin.* Preface by Eli Wiesel. Translated by Sara Hughes. Watertown: Blue Crane Books.

Zürcher, Erik J. 1998. *Turkey: A Modern History.* New York: I. B. Tauris.

7

STRUCTURES OF
A REVOLUTIONARY PEDAGOGY

PALESTINIAN POLITICAL CAPTIVES
IN ISRAELI PRISONS

Esmail Nashif

One of the major sites of the Palestinian national movement is the Israeli jail.[1] In this site of intense and tense colonial relations the prison, the pedagogy, and the revolution are interwoven to create a revolutionary Palestinian pedagogical system. The far-reaching effects of this pedagogical system on the lives of the Palestinians, and by extension of the Israelis, are the main focus of this chapter.

Hasan Abdallah,[2] a former political captive, who is the main interviewee in this chapter, describes his first meeting with this pedagogical system as a shocking but a constitutive one:

> When I entered the prison for the first time I was so surprised. I had come from the university. And you know when somebody comes from the university he thinks he knows something, he feels that he knows more than the others. This is because he can read any book and so on . . . I was surprised that after a month I had to change and deal with critical cultural issues, in order to benefit from them [the political captives] . . . so I joined in their discussions, which I felt were far more sophisticated than the university's . . . . (Abdallah interview, September 6, 2001)

The rebuilding of the Palestinian communities of the West Bank and the Gaza Strip, from shattered populations living under occupation into the "revolutionary society of the Intifada," in the words of al Dyk (1993), depended on the creation and transformation of institutions that could sustain this resistance. Palestinian intellectual Lisa Taraki, in a 1990 article, described how universities, unions, voluntary groups, and other associations were transformed by mass participation in these joint endeavors as well as by the symbolic meanings attached to them. While Taraki did not examine Israeli prisons in any detail, she ended her essay with the recognition that they too "played a significant role in the politicization of Palestinians, especially the youth":

> Israeli jails, a powerful symbol in the political folklore of the occupied territories, are often considered by their "graduates" as the ideal place for acquiring a political education. There, isolated from the routine of normal life, prisoners organize seminars and study circles, conduct Hebrew and English lessons, and teach the illiterate among them how to read and write. The impact of this collective experience is no doubt a lasting one. (Taraki 1990, p. 68)[3]

This chapter explores the "impact of this collective experience" by examining the complexities, contradictions, and conflicts of being a Palestinian political captive in the Israeli prison system.

According to surveys by local and international human rights organizations, at least one-quarter of the Palestinian society has been imprisoned by Israeli authorities at some time. For example, the Amnesty International report for 1993 stated that 813,000 Palestinians out of three million living in the occupied territories had been imprisoned since 1967 (cited in al Hindi 2000). Since the late 1980s, the annual Amnesty International report on human rights conditions in Israel and the occupied territories has regularly begun with the issue of political imprisonment. The report of 1989 opens with the following:

> About 25,000 Palestinians, including prisoners of conscience, were arrested in connection with the *intifada* (uprising) in the Occupied Territories. Over 4,000 served periods in administrative detention without charge or trial. Several thousands others were tried by military courts. By the end of the year over 13,000 people were still in prisons or detention centers. (Amnesty International 1990, p. 129)

These mass arrests had major effects on the Palestinian society in the occupied territories, even on a daily basis. By regularly raising issues of political captivity since the mid-1970s, the Palestinian media brought the issue to the attention of Palestinian society as part of the national daily agenda. When hunger strikes by political captives inspired Palestinians outside the prisons to protest and demonstrate, the effect of the lessons learned by the captives spread far beyond the prison walls. Thus, while political captivity was an expected result of political activism, it then spread back to reach wider social circles. Specifically, the centrality of the captivity experience in reshaping the Palestinian identities and ideologies in the colonial system is manifested in the political, social, and cultural aspects of the Palestinian national movement. This is reflected in the fact that in almost every political speech of any Palestinian leader, the act of turning the colonial jail into a "revolutionary school" for the cadres of his/her political organization is celebrated as a victory. Furthermore, the values of cooperation, solidarity, and equality, which are said to reign in the community of political captives, are propagated as a social model for Palestinian society in general. On the cultural level, the powerful, almost mythical, symbol of the Palestinian captive who rises from the ashes of the colonial jail into history by educated resistance is a recurring trope in Palestinian literature, art, and daily storytelling among political activists and laymen.

This chapter's exploration of the complex social field of the "prison" in general, and of Palestinian political imprisonment in particular, begins by setting aside two opposing meanings, though one structure, of "othering": criminality and romance. The Israeli metadiscourse, in the academic arena as well as in other spheres, sees the political captives as terrorists, thus criminalizing their practices before, during, and after imprisonment. The Palestinian mainstream discourse transforms them into mythical heroes. Both these attributions—of criminality and of heroism—conceal, distort, and deny the experience of captivity. Imprisonment is interrogation, solitary confinement, torture, life sentences, and death. At all times, being a prisoner means dealing with oppression and engaging in resistance, which transforms prisons into structures where revolutionary practices have to be learned. By uncovering these naked power relations between the colonizer and the colonized, this chapter focuses on the (re)production of a new field of meaning through pedagogical practices. In this context, pedagogy is seen as a prerequisite for a true revolution.

Many former Palestinian political captives have written about their experience with political imprisonment, and others have described the collective experience in interviews conducted after their release in the early 1990s.[4] Abdulstar Qasim and his students (1986) at Al Najah University in Nablus, most of them former political captives, wrote one of the first historical studies on political imprisonment for the Palestinian public. Abdallah (1994, 1996), who has been arrested several times since the early 1980s, explores the literary history of political captivity. Since his release from his second term of political imprisonment, Fahid Abu il Haj (1992) has been collecting and publishing the life stories of political captives. Together with other former political captives and political activists, Abu il Haj established The Abu Jihad Center for the Political Captives Movement. The center, which is located in Ramallah, engages in documenting and collecting the texts, artistic creations, and narratives produced by the political captives. These are some of the examples of the enormous corpus of written materials about the Palestinian political captivity in the Israeli colonial jail. Many of these texts were written while the authors were political captives serving their term. In the interviews and in the histories that they have written the former political captives identified the practice of reading/writing and the prominent place of textual interpretations and (re)productions in the fabric of the political captives' community as the main achievement of the political captives' movement. The present chapter attempts to explore historically and socially the pedagogical mechanisms of this community. In the context of the national conflict, the assumption that this community could not sustain itself without its own pedagogy leads to an imperative need to explore its social and historical premises. Seen from this perspective, pedagogy is defined as the intentionally formalized practices of a community to sustain and reproduce itself, while disseminating its dominant ideology to the different individuals who make it.

The harsh conditions of their imprisonment, which were intended by the Israeli authorities to resocialize Palestinians into docile and submissive bodies/souls, left the political captives with meanings that were as ruptured as the colonial prison space/time. The discourse of *thaqafah* (culture) became the site for captives to resist the effect of the prison by constructing, through the praxis of writing/reading, a counter-hegemonic symbolic and material field of action. The creation, dissemination, and propagation of *thaqafah* as a space between captives that transcended the space of the prisons thus became the empowered site for the Palestinians' revolutionary pedagogy.

This experience of the Palestinian political captives with *thaqafah* provides a new perspective on Michel Foucault's argument about "the technologies of the soul" and the development of the modern prison. For Foucault, the subject of study is not the prisoner or the prison, but rather the technologies of power:

> What was at issue was not whether the prison environment was too harsh or too aseptic, too punitive or too efficient, but its materiality as an instrument and vector of power; it is this whole technology of power over the body that the technology of the "soul"—that of educationalists, psychologists, and psychiatrists—fails to conceal or to compensate, for the simple reason that it is one of its tools. (Foucault 1978, p. 30)

Analysis of the prison's "materiality as an instrument and vector of power" shifts the interpretive framework from particular conditions to the underlying power relations that the prison in totality enforces upon and through the bodies and souls of prisoners. Following Foucault, this chapter asks how the submissive and helpless position of Palestinian captives is secured and sustained through "the technology of power over the body" and "the technology of the soul" that are more openly employed in prisons than in other "correctional" facilities such as schools, hospitals, and factories.

But this chapter also goes beyond Foucault to examine the agency of prisoners and the possibilities of a counter-hegemonic discourse articulated by imprisoned subjects. This argument rests upon the assumption that in any sociohistorical context of power relations, certain spaces/times of resistance coexist concurrently with oppression and dominance. In the case of the Palestinian political captivity, reading/writing became the praxis of resistance to the Israeli colonial jailing system, not just in and by itself, but more importantly as part of the community building process.[5] It is important to emphasize at this stage that I use the term community to designate all the Palestinian political captives in all the Israeli prisons, who were part of the Palestine Liberation Organization's (PLO) various organizations.[6] Prominent among these organizations are the mainstream Fatah and the Marxist-Leninist Popular Front for the Liberation of Palestine (PFLP).

THE PEDAGOGICAL FUNCTIONS OF THE JAIL

The community of Palestinian political captives uses this meaning of culture, *thaqafah*, to define the whole ideological apparatus through which the community seeks to reinstall, reconstitute, and reaffirm "Palestinian-ness" as a national identity. In fact, in this context *thaqafah* delimits a site liberated from prison conditions and thus liberating for the captives, with the ultimate goal of making the lessons learned in prison an instrument of liberation for all Palestinians.[7]

This community is not a set of relationships brought from the outside, though it is linked to the outside world. It is a community that was constituted mainly by the processes of struggling with the spatiotemporal grids of the colonial jailing system, a struggle that provided the community with new options for resistance. Prominent among these options are the writing and rereading of the community's own identities and ideologies in a historical perspective. The almost formal (in the sociological sense) pedagogical institutions of the political captives community inculcate in the individual political captive a structure of feeling that we can call "the historical mission." The political captive is defined as part of an irreversible historical movement, which will lead eventually to the liberation of Palestine. The complexity of the relationship between pedagogy and history in this community not only helps us to understand the larger workings of these relationships in Palestinian society, but also casts light in general on the relationship between pedagogy, as an ideological apparatus, and history, as a context partly determined by objective conditions of materiality.

Both the prison and the captive community are materialized in the bodies of the captives, the former as a means of control and the latter as a weapon of resistance. In his study of the processes of violence formation in North Ireland, Allen Feldman explores the political agency and the body, as an agent/object of violence. He interviewed former political prisoners, Catholics and Protestants, Nationalists and Loyalists. His ethnographic data led him to narrate the events of violence through excavating the narratives of violence as told by the interviewees. In his attempt to understand the body in relation to time and space in modernity, and building on Foucault's interventions, he argues that:

> The political form and the commodity form fuse because, in modernity, political power increasingly becomes a matter of regimenting the circulation of bodies in time and space in a manner analogous to the circulation of things. Power, as Foucault amply documented, becomes spatialized. (Feldman 1991, p. 8)

The Israeli prison system is a modern surveillance system par excellence, especially if we notice that it was erected, literally and metaphorically, on the British jailing system in mandatory Palestine. Locating the relations between the Palestinian political captives and their Israeli captors in modernity enables us to understand the mechanisms of circulating the Palestinian political bodies in the severely regimented squares of time and space in the occupied territories. Feldman, based on his fieldwork, argues that reducing the space allocated to the body under surveillance interacts directly with the perceived agency of the resisting subject. He claims ". . . the shrinkage of the space of political enactment corresponds to the expansion of the acting subject—the increasing correlation of personhood to historical transformation" (p. 10). In the Palestinian case, this dialectic of resisting the spatially defined limits of the colonizers by expanding the agency of the colonized is realized by and through the categories of narration used by the Palestinian political prisoners to characterize the differences among diverse Israeli prisons. Those prisons that are populated by prisoners with long sentences are said to be the most communally organized, intellectually elaborate, culturally democratic, and politically sophisticated. The history of 'Sqalan prison is an example of these processes of spatial shrinkage and expansion of the agency of the political captive. Until the late 1970s and early 1980s, the shortest sentence was 15 years. Beating and abuse of the captives was a daily occurrence. The prison authority systematically deprived the captives of everything they managed to achieve by strikes and protests. In spite of all this, or more accurately, in direct relation to these harsh measures, the captive community of 'Sqalan came to take the leading role in the entire captive community in Palestine.[8]

The community of Palestinian political captives is characterized by continuous confrontations, mediations, and negotiations between the captives and the prison authorities. These interactions gradually became regulated and institutionalized in specific communal functional structures. Prominent among these functional structures are the ones designed to (re)produce knowledge and power inside the community and vis-à-vis the prison authorities. One of the main arguments of this chapter is that the captives, their community, and *thaqafah* constitute the agents, the location, and the products, respectively, of a revolutionary pedagogy. Moreover, the unique educational system built by the political captives came to be one of the main dynamic sites for reproducing the Palestinian subject within the Palestinian community.

Louis Althusser's theoretical insights about ideological state apparatuses provide another perspective on the Palestinian political captives community as a self-constituted educational system. Althusser distinguishes between ideological state apparatuses (ISA) and repressive state apparatuses (RSA) by arguing that the former use ideology as a representation and as a materiality, while the latter use violence:

> The (Repressive) State Apparatus functions massively and predominantly by *repression* (including physical repression), while functioning secondarily by ideology. . . . In the same way, but inversely, it is essential to say that for their part the Ideological State Apparatuses function massively and predominantly by *ideology*, but they also function secondarily by repression, even if ultimately, but only ultimately, this is very attenuated and concealed, even symbolic. (Althusser 1971, p. 145; italic in original)

Moreover, Althusser argues that the educational ISA is the most dominant among ISAs in the age of mature capitalism, which includes the unique circumstances of modernity as embodied in the colonial relations between Israelis and Palestinians in the occupied territories.

The negotiations, mediations, and confrontations between the Palestinians and the Israelis in the context of the colonial jail are constellations of dynamic relations. At least formally, the Israelis control the physical and the material, aiming to redefine the Palestinian subject. The Palestinians, on their side, control the meanings of being a subject in order to seize the physical and the material. At this juncture of the colonial relations, the production of meaning/knowledge becomes crucial for both sides. Hence, the political captive community's educational system is seen as a strategy for redefining the hybrid meanings of the colonial context into an alternative, resisting meaning in order to liberate the material, namely the land of Palestine, and to build on it a nation-state. This educational system emerged through the dynamics of negotiating and contesting the already existing pedagogical systems in the Palestinian society, and the ones imposed by the Israeli authorities.[9]

The social formation of the Israeli jail encompasses a complex set of interrelations between the repressive and the ideological, and between the state apparatuses and the community apparatuses. The colonizer–colonized class division is materialized both metaphorically and literally in these ideological–repressive formations. The actions, practices, and rituals of the ideological formation of the colonial prison, that is to say, the revolutionary educational system of the Palestinian political captives, the colonial authorities and their interrelations, contain at the same historical moment of colonizer–colonized, the ideological conditions for reproducing and transforming the colonial relations as the dominant mode of production. In this chapter, I try to show that the Palestinian political captives community has transformed these colonial relations through reframing them as changeable via resistance, and hence putting the two sides of the conflict, at least, on the same footing.

As for the concrete ways by which these ideological apparatuses work, Althusser positions the real subject in the center of the apparatus, by the actions, practices, and rituals that s/he performs. In fact, Althusser's main argument is that the apparatus exists only through the praxes of a subject:

> . . . when a single subject . . . is concerned, the existence of the ideas of his belief is material in that *his ideas are his material actions inserted into material practices governed by material rituals which are themselves defined by the material ideological apparatus from which derive the ideas of that subject.* (Althusser 1971, p. 169; italic in original)

Thus, the actions, practices, and rituals constitute the basic elements by which ideological apparatuses are built and crystallized. From this point of view, in order to understand the educational system of the captives' community, one must identify the actions and practices of the individual political captive as a subject located in these apparatuses.

To demonstrate these processes of producing and reproducing meaning through the actions, practices, and rituals of the Palestinian political captives, I present the case of a former political captive, Abdallah who wrote extensively on the issue of political imprisonment while he was a captive, and he continues to write using different genres. Abdallah's voice is unique in the problematic that it raises in the context of the period following the Oslo Accords, which were supposed to put an end to the Palestinian–Israeli conflict and hence to the issues of political captivity. In a sense, he is the historian of the era of political imprisonment, which was dismissed as part of the ancient regime that had to be forgotten. Being on the borders of these different Palestinian national historical constellations, Abdallah, through his actions and practices, helps us to understand the transformations in the Palestinians' colonial experience with its many complicated facets.

The Palestinian captives community in the Israeli jail is a complex multilayered sociohistorical situation. In order to understand it in its many facets, I use Foucault's understanding of the prison system, on the one hand, and Althusser's intervention about the dynamics of ideological apparatuses, on the other. The first section addresses the question of the historical positionality of political captivity in Palestinian society through the community members' practice of reading/writing. The second section presents, through the case of Hasan Abdallah, the specific actions, practices, and rituals that constitute the ideological apparatus of this captive community as a revolutionary pedagogy. In the third part of the chapter I try to reconnect these practices with the Arab Palestinian society through its different formations and structures. Understanding the captive community as existing in a state of liminality (Bhabha 1990a) is one way to analyze the pedagogy of revolution that was developed by the political captives.

Finally, the questions and issues invoked by the case of the Palestinian political captives on the practical and theoretical levels are relevant both for an understanding of the constitutive dynamics of the national conflict between the Palestinians and the Israelis and for opening new horizons to explore the human conditions in which hierarchical power structures construe the deep schisms between the material and social conditions of existence. Thus, the ideological (mis)recognitions about the separate domains of the material and the intellectual explorations could be transcended. The illuminating example of the Palestinian political captives' educational practices, seen in its historical context and sociopolitical processes, indicates the thin lines that blind social and historical inquiries to the Gramscian insight that each social group, based in objective material conditions, creates its intellectuals in order to survive.

WRITING THE HISTORY OF THE PRISONS

The Palestinian national movement and political consciousness developed, as Taraki shows us, by capitalizing on modernity's logic of affiliation, namely, modern institutions such as political parties and professional associations (Taraki 1990). However, these relations of communitas are not fixed, but rather contingent on the specific historicity of Palestinian society. The history of this society has at its disposal many social formations, which manifest and materialize on the cultural level more than in any other social sphere as one of the characteristics of the Palestinian locality of colonization. The colonization processes shattered the nuclei of the competing agricultural, mercantile, and dependent small-scale industrial modes of production (Saleh 1990).

Despite these intense processes of transforming the infrastructure of Palestinian society, the cultural systems and practices of significations that correspond to the shattered socioeconomic patterns of activity persist in different ways. For our purposes here, the Palestinian national identities and ideologies dominate, compete, and merge largely by articulating these histories of the many social formations. In the context of the colonial jail, the Palestinian city dweller, the former landlord, the peasant, the manual worker, and the professional, among others, meet in the confined colonial space/time. This meeting created a community, a subject, but this new community has certainly not replaced the preexisting identities, rather it interacts, competes, and merges with them (Bhabha 1990a). How does the community of captives as a whole interact with these pasts? Do the captives rely on specific cultural patterns of behavior to cope with the new situation of captivity? And mainly for our purpose here, how did the reading/writing practices, as a revolutionary pedagogy, come to be the prominent dynamic of producing knowledge/power and thereby producing and reproducing the community of captives?

For Palestinian captives, writing in and out of prison about the history of the prison became an act of writing the history of the present, as Foucault sought to do in *Discipline and Punish* (Foucault 1978). In an interview with Muhammad 'layan, a former political captive and a short-story writer, I asked:

> What did it mean for you, literary writing in the prison? I mean in general and practically, how did you do it?

Muhammad answered in an affirmative and authoritative voice:

> There is no such a thing as literary writing; in the prison everybody writes . . . the different genres are less important . . . you write yourself, you don't have many other options . . . so you write all the time even if you don't have the tools for writing. ('layan interview, December 6, 2001)

The reading/writing sign, site, and space of actions, practices, and rituals are still the main activities practiced by the political captives in order to establish themselves as a social group. As Palestinian literature and interviews with former captives show us, learning, through reading/writing, was liberated through struggle against the prison authorities. Hence, the question arises: how can we deconstruct the social space of the reading/writing activities, on the level of both material and social relations? For if we understand the relational elements of the social space created by the reading/writing activities, then we can better understand the ways in which this social group (re)built itself through its revolutionary pedagogy.

The broader issues that deconstruction of the social space of reading/writing may help us to explore are the traces of other Palestinian social formations, which are articulated and repositioned by the different sections of the Palestinian national movement. If education was confined to certain social groups in the near past, it could not be so in the new modern community of the political captives or the larger Palestinian society. It goes without saying that this was not a linear modernization, on the contrary it created diverse and alternative shapes of modernity, largely due to the absence of an independent Palestinian nation-state as the main actor in a stable hegemony (Doumani 1995; Kimmerling and Migdal 1993; Swedenburg 1995). Thus, by tracing back the practices of revolutionary pedagogy as equally accessible and widespread among most of the Palestinian political captives, we may locate these shapes of Palestinian modernity as a historical continuum of construction of the Palestinian community.

This section focuses on Abdallah's historical account, as narrated in interviews, and uses it to trace the history of reading/writing in the Israeli colonial jail. In addition to Abdallah's account, I introduce other former captives' narratives in order to highlight and also to criticize some aspects of the main narrative. As mentioned above, Abdallah's account of the political captives' educational system is part of his larger written corpus documenting the literary history of the political captivity experience. As such, his narration is self-conscious and locates itself as an intervention designed to shape Palestinian society and counter the Israeli narrative about its history. Hence, beyond providing insights about the ways in which the history of the political captives' experience is told, Abdallah's account may cast light on the way in which Palestinian society at large retells its own history.

HASAN ABDALLAH

The name of Hasan Abdallah is familiar to anyone interested in the subject of Palestinian political captives. When I began my fieldwork in the summer of 2000, I made inquiries through formal and informal channels and especially in the literary circles in Ramallah, about former political captives who had published literary texts written while imprisoned by the Israelis. Abdallah's name was mentioned most frequently, not only as a former political captive, but also, and in ways that turned out even more important for my fieldwork, as the local historian/critic of the prison literary (re)production. His written corpus introduced me to the names, titles, and dynamics of political imprisonment.[10] Meeting Abdallah was my initiation into the textual realities of Palestinian political imprisonment.

My first interview with Abdallah took place in Ramallah on September 6, 2001. It was in his office at *Watan* TV, where he works as chief news editor. Those were the stormy days of the current Palestinian Intifada (before the September 11 attack in America), when political and military events and operations in the occupied territories were occurring on an hourly basis. The interview was interrupted many times by reporters and journalists, coming with items of information and updated events. The flow of telling and listening was extremely discontinuous:

> I am one of those people who were arrested while studying at the university. I was imprisoned from 1981–83, and spent those two years in 'Sqalan. This experience was so important to me that it superseded the university experience. To speak about this experience, we must talk about the prison . . . 'Sqalan . . . 'Sqalan prison was opened after the occupation of the West Bank and the Gaza Strip. And its purpose was to resocialize the Palestinian fida'yyn [freedom fighters], those who came as infiltrators from abroad and the Palestinians who were sentenced for long terms. For 15 years no new prisoners with short sentences came to 'Sqalan, only those with long sentences. In 1981 I met prisoners who had not left 'Sqalan since 1967, not even for transfer [to another prison] or medical treatment. 'Sqalan prison was closed, closed . . . This closed experience was an exciting one. I felt that I was entering a closed city . . . like that . . . it had its norms, habits . . . it had conventions, the people talked about a temporal context, reminding each other of "stations" [landmark events that changed the conditions of imprisonment]. This means that their memory had become confined in the prison to the prison's history . . . the workings of the prison . . . how it developed . . . the stations of struggle that the prison had gone through . . . the cultural, organizational, and struggle stations. . . . When this prison opened, it was a punitive one . . . in the morning the Palestinian prisoner had to eat the beating meal of the morning . . . and there was the evening beating meal. . . . (Abdallah interview, September 6, 2001)[11]

Abdallah lived the memory of specific landmark events that had changed the relations between the prison authority and the captive community as "stations" in the developing history of the political captives' community. These historical stations were recounted and articulated to him as a novice in 'Sqalan by the older political captives. Each moment of confrontation with the prison authorities was transmitted to him in the ritual of retelling histories of the prison to transform him into a member of the community. In his *The Literary Production of Political Imprisonment*, he reiterates the stations textually:

> To start to study and analyze the prison literature, one must deal with the living, social, intellectual and psychological conditions that formed the first years of the lives of the first political prisoners, and continued to influence them and the thousands that followed . . . The beginnings especially formed the basis, which brought a more mature and organized stage of political imprisonment. . . . (Abdallah 1994, p. 13)

Abdallah narrates the history of the captives as a community based on heroic beginnings. At each beginning stage, there are two poles, two antitheses that are asymmetrical in their power relations. One is a human struggling to survive, while the other is a brutal colonizer, who does everything to eliminate "the Palestinian." The more the body, its space and time, suffer, that is, the more it is subjected to harsh measures of centralized spatiotemporal domination at the hands of the colonizer, the stronger and deeper becomes its cohesion as a community.

This dynamic of relations between the Israeli and the Palestinian in the context of political captivity elaborates some aspects of the two theoretical tracks we took in the introduction. First, the idea of state apparatuses in a nation-state context in Althusser's argumentation could be extended and examined in the context of a national community, such as the Palestinian community, that does not possess a modern nation-state bureaucratic system. The critical question seems to be the ways in which the dominant ideologies are articulated and materialized by different mechanisms. Second, these dynamics seem to follow the logic of spatializing the power relations, as described by Feldman in discussing the case of Northern Ireland; that is to say, the harsher the measures of confining and regimenting the captive's body/soul, the more politically active the latter will be.

The centrality of reading/writing and education in general combine in the oral and written narratives of the former political captives. For example, Faris Qadwrah, who is an elected member of the Palestinian Legislative Council, describes his experience with almost a spirit of emancipation:

> I was from the third generation of political captives . . . I was not educated before I was imprisoned . . . in fact I was a simple worker . . . I was assisting a blacksmith . . . Then in prison I had a big chance and I used it . . . I had some basic skills in Arabic, but it was not much of a help, and I couldn't stay ignorant . . . so I took courses in Arabic and in other subjects. (Qadwrah interview, February 17, 2002)

The transformational experience of political captivity for Qadwrah was not captivity itself. Rather, the perception of oneself changed through acquiring skills of producing knowledge. While Qadwrah saw in knowledge its instrumentality, others were enthusiastic about knowledge in the broader issues that producing knowledge could give them.

Ata al Qaymari, a critical and articulate journalist, was imprisoned for 14 years, from 1971 to 1985. During the last two years of his captivity, he started to write the history of the political captivity of the Palestinians since 1967. In *al Sijnu Laysa Lana*, translated as *The Prison is Not for Us*, he describes the dynamics of confiscation by the prison authorities of any object connected with cultural activities. For example, pen, paper, and books were forbidden until 1971. The political captives' counteractivities to this policy of confiscation, literally and metaphorically, were constant struggles to regain the confiscated objects (al Qaymari 1984, pp. 117–125). After his release, al Qaymari opened a translation and press service. When I met him for the interview, he was still running the service from his office in Jerusalem.

> Then they moved me to the central prison in Ramlih where they put all the Palestinian political prisoners from Jerusalem . . . One day a French prisoner came . . . he had almost 2000 books, but they were in French, I made a deal with him . . . cigarettes for books . . . What was forbidden in Arabic was available in French . . . I asked my organizational superior [for permission] to study French . . . He refused saying "you don't have time for it" . . . after a while he

agreed to an hour a day . . . The first book I translated was that of [Vo Nguyen] Giap, the Vietnamese general, about the popular war . . . The superior was astonished, then we made several handwritten copies of it and distributed them among the prisoners. (al Qaymari interview, January 13, 2002)

In contrast to Qadwrah, al Qaymari's narrative indicates the centrality of reading/writing in the political captive community not as a general amorphous practice, but rather as a specific kind of activity that is integral to the hegemonic ideology of each Palestinian organization. Although each community in the various colonial jails has general activities, such as lectures and courses for all the political captives, most of the cultural activities are conducted by each organization independently of the others.

A year after his release from political imprisonment in 1987, Muhammad Lutfy Khalyl wrote a book about the dynamics of the community of political captives, in which he argues that the cultural activities of the community were institutionalized according to the organizational divisions:

> In fact there is no one cultural system in the prison, but each faction [political organization] has its own system, especially the big factions. Each cultural system starts with a higher committee of three members, who are known for their high cultural abilities and especially their understanding of the faction's program, political principles, positions and aims . . . The members choose a president [for the committee] who is usually a member of the general central committee of the faction . . . this procedure is, in most of the cases, in consultation with the [political] leadership . . . Then the committee appoints a cultural officer on the section level and one on the room level, too. (Khalyl 1989, p. 108)

The ideological community apparatus has many mechanisms of interpellating the Palestinian as a subject. For example, the political captives community uses the Palestinian national calendar, which divides the time into cycles that reinstall, among other things, the time frame of the Palestinian national identity.[12] But mainly, this community of Palestinian political captives uses systematic educational activities, termed *thaqafah*, to interpellate its subject, basically along the different organizational ideological lines. The word *thaqafah*, which literally means culture, denotes the whole ideological apparatus through which the community sought to reinstall, reconstitute, and reaffirm "Palestinian-ness" as a national identity and ideology. These main objectives were gained through institutionalized cultural practices centered around reading/writing as the core of *thaqafah*. Moreover, the systematic educational activities, while building an identity, were designed to counteract the hegemonic colonial practices of annulling the Palestinian subject.

During his second term of imprisonment, from 1985 to 1988, in Jnayd prison in Nablus, Abdallah became the librarian of the captive community.[13] Positioning the *thaqafah* in the context of struggle against colonizers' attempts to redefine the Palestinian subject, Abdallah explains:

> The political prisoners, from the beginning, understood the importance of the cultural side, and they understood that the prison authorities were trying to

empty the Palestinian prisoner or struggler of his cultural content; and when you empty him of his cultural content, whether a prisoner or not a prisoner, it will be easy to make him docile, to break him. Then it will be easier to fill his mind with other ideas . . . The prison authorities wanted to turn the prison into a cultural wasteland for us. (Abdallah interview, September 6, 2001)

The colonial authorities' constant attempts to empty the Palestinian captive recur in texts written in prison and outside. But these attempts are always countered by the *thaqafah* practices of the political captives' community, as narrated in the interviews and texts (Abdallah 1994; Khalyl 1989; Qaraqi 2001; Qasim 1986; al Qaymari 1984). Reading/writing was established, as the action of inscribing the national ideology and expanding the persona, which would transform history. Through different stations of confrontations and struggles, the pen, the paper, and the book became some of the basic signifying materialities of the community's identity (Qasim 1986).

The basic action of the *thaqafah* is the reading/writing activities and skills. The practice of these actions follows a regular schedule: daily, usually twice daily, study circles, reading sessions, lectures at the level of the room and the section, lessons on scholastic topics and training in professional skills, literary/intellectual group discussions, oral transmission of the local history during the daily break, and physical exercises. Alongside these organized activities, many of the political captives practice writing and reading as part of their organizational duties (communiqués and articles), and some of them engage in literary reading and writing activities using different genres of literature. Thus, the basic skills of reading, writing, and interpretation must be seen and positioned in the larger context of the processes of constituting a resisting community. It is worth noting that these communities of the Palestinian national movement are secular modern ones, unlike the Palestinian Islamic movements that developed in a later phase of the Israeli–Palestinian conflict, especially in the late 1980s.

At the other end of this range of practices, there were many study groups on intellectual and literary topics, which transcended the narrow organizational affiliations:

I participated in many of what could be called conferences on novels by Hana Mynah[14] . . . I remember that in 'Sqalan there was a novel by John Steinbeck, I forget the exact title, *The Ghost City* . . . or something like that . . . It was so popular among the prisoners . . . that I met people who had read it ten times, and the paper faded and you couldn't see the words . . . So we kept rewriting it all the time, making new copies for the prisoners to read. (Abdallah interview, September 15, 2001)

With the passing of time and the accumulation of skills and knowledge, the division of labor in the community became more differentiated. The growing differentiation and specialization called for the establishment of social rituals granting recognition to certain individuals entering the public sphere of the community as experts:

> There were many experts on different topics. Faris Qadwrah was famous for his foreign languages skills; Jibryl Rajwb was an expert on Israeli studies. Others were experts on political economy . . . There was one Hafiz 'byat who was the most expert person on "The Origin of the Family, Private Property, and the State" by Engels. I took a course on the book with this man, for a whole course we discussed it. (Abdallah interview, September 15, 2001)

The processes leading up to the public initiation of an expert are punctuated by stations marking the accomplishment of certain reading/writing practices, with specific social values gained at each station. It could be argued that in the organizational and interorganizational contexts of political captivity, reading a certain corpus of literature and writing certain texts qualify the captive as an expert on a certain subject:[15]

> I was known for literary and journalistic skills . . . When someone new comes to the prison, and is known for being interested in such and such, people direct him to prisoners who have the same interests. (Abdallah interview, September 15, 2001)

The basic skills, practices, and social rituals, together with the elaboration of a differentiated division of labor, defined the borders of the community, spatially and temporally. The mechanisms of inclusion in and exclusion from the community, with the elaborated skills and practices of reading/writing, took clearer and more coherent shape.

The next stage in these processes of community structuration was the building of networks among the different colonial jails. An example of the operation of this communication network is the coordinated measures taken by the captives in the different prisons, such as strike days. Thus, the borders of the political captive community stretched to include all of the Palestinian political captives, transcending boundaries between prisons. Three main factors enabled the captive community to enlarge and become more dynamic: the consolidation of the PLO's social and political networks in all the occupied territories, the sophisticated postal system of the captives themselves, and the more flexible formal communication circuits with the outside world, such as newspapers and radio. These circuits of information and knowledge work in both ways, from the larger society to the colonial jail, and from the latter to the daily lives of the Palestinians in their social settings. These mutual relations, especially when all of the political captives started to coordinate their activities as one group, had some passing effects on the Palestinian national movement, while other effects were more lasting ones. This section thus asks how these new relations interact, conflictually or otherwise, with the social formations of Palestinian society at large? More specifically, how does this revolutionary educational ideological community apparatus interact with the educational system outside the borders of the captive community?

The actions, practices, and rituals of the prison community are constantly compared with the parallel ones of the larger Palestinian community in the occupied territories, based on each organization's view of Palestinian society. There are two main dynamics of comparison. First, Fatah, the largest mainstream Palestinian organization, is based on a nationalist ideology and seeks to liberate Palestine by subordinating the different social formations to the organized effort to build a nation-state. This does not necessarily mean changing Palestinian society as long as these social formations are compatible with the efforts aimed at national liberation. The second is that of the Palestinian left, mainly represented by the PFLP. The Palestinian left sees the different Palestinian social formations as an obstacle, another front to deal with in addition to the occupation. Their solution is to disseminate an alternative *thaqafah*.[16]

Abdallah retells the experience by comparing the left with Fatah:

> Fatah was concerned with the organizational side, how to attract new members. They were concerned mainly with the national [Palestinian] culture; this is in general. The left was investing a lot in the cultural dimensions and this I am telling you objectively. Always the left led the thaqafah in the prison. . . . (Abdallah interview, September 22, 2001)

While Abdallah tried at one stage of the interview to attribute the complex relations with the larger society to differences in the organizations' ideologies, at another point in the interview he raised new dimensions. This use of multiple and contradictory strategies of narration by Abdallah characterizes many former political prisoners:

> . . . it does not mean that in all the organizations there were things forbidden; it depends on the people who are in charge. This means that a certain leader at a specific time could be a democratic person or an oppressive one, could have problems with certain social issues or not; it could be that the woman issue is an obsession for him, and he wants to suppress it in all the people under him. He wants to marginalize it . . . In the prison personality problems were reflected in the experience [of political imprisonment] . . . . (Abdallah interview, September 22, 2001)

Abdallah continues describing and analyzing these problems by relating and connecting them to the larger Palestinian social divisions:

> . . . and there is a big difference between somebody who finished a year in the university and someone who didn't. If I come from an open [liberal] social environment, it will not be easy to oppress me and to pattern me in a frame. In contrast, one who is 16–17 years old, a newcomer from a closed [conservative] family, from a village with a limited range of relations . . . It made a big difference if you were from the city, or from the countryside . . . and it made itself felt in the prison, the camp [refugee camp] and there were different social coalitions that intervened and affected the organizational life of the prison. . . . (Abdallah interview, September 22, 2001)

The socioeconomic and the personal histories interplayed with the objective conditions of captivity to form the organizational sides of the community. Since the War of 1967, these socioeconomic structures were drastically transformed (Hiltermann 1991).

The occupation of Palestinian society in the West Bank and Gaza strip, sometimes intentionally and sometimes as a by-product, generated processes of social and economic transformation, which led to the destruction of the existing socioeconomic formations, such as the agricultural and small-scale industrial domains. Yet, the social and cultural differences persisted and were part of the new identities and ideologies in the Palestinian national movement, and in that capacity played a major role in the specific constitutive processes of the political captive community.[17] The materiality of these social formations was articulated mainly in ideological forms and practices of prohibitions and borders of inclusion/exclusion, not only of individuals, but also of social spheres of activity for these individuals, as in Abdallah's account:

> . . . for example I saw the censor [appointed by the Palestinian organizations] in the prison . . . as you know several Hebrew papers were sent to us, one of them was Ha'olam Hazeh,[18] and in this magazine there were a lot of scenes and photos of almost full nudity. Now the censor comes and surveys it. He has ink, so he covers the photos with the ink, so that when the magazine is read by the different prisoners nobody will see it . . . Of course who sees it? The censor . . . ohhh . . . the censor sees what the others do not see. (Abdallah interview, September 22, 2001)

Such prohibitions and practices of exclusion/inclusion were institutionalized in the political captives' community by the appointment of a censor for each organization, and by formal prohibition delivered in written orders to the cadres.

In the matter of sexuality among the captive community, the ideological and repressive apparatuses merge at different levels. Censorship is epistemic and physical violence. In the context of sexuality, we see how the community domesticates systems of signification from the larger Palestinian society, in this instance patriarchy, to appropriate and to control the "body." In this way, the political captives' experience reproduces the general social and cultural practices, and does not transform them. Thus, the multiple practices of prohibition are signification practices that mark both the outer borders of the community and the inner ones that demarcate individuals in their relations with the community.

Still, some issues that are inherent in the patriarchal structures of society seem to have more valued meanings than others do. This positionality led to overinvestment in implementing prohibitions related to these social domains. Such is the case with sexuality. The prison of sexuality, if one may say so, is a kind of overdetermination. It recurs in different domains of the social and cultural spheres with different manifestations but with the same deep structure. Thus, we see the different expressions of the dynamics of prohibition practiced by the censor, the father, and the political leader. The intense relations in the prison add the colonizers' practices to these patterns of signification.

The Palestinian practices of prohibition interact and merge into a unique kind of hybridity with the colonial practices of prohibition. The mass media in the occupied Palestinian territories of the West Bank and the Gaza Strip can serve to illustrate this argument. The Israeli colonial authorities forbade any expression, whether written, visual, or auditory, that had anything to do with Palestine, Palestinian national history or national identity/ideology. Like every other infrastructure, the Palestinian press was curtailed by systematic Israeli orders and actions (Farsoun and Landis 1990). During the era of occupation from 1967 to 1993, there were no Palestinian radio or TV stations, only several heavily censored newspapers. Parallel with the colonial prohibitions, the traditional Palestinian social mechanisms of enforcing prohibitions were weakened and redefined by the rising national ideologies.

The political captive community is a dynamic example of these processes. Although the political captives developed their own communication systems, the flow of information was heavily censored by the prison authorities. For example, radio and TV sets were forbidden until the mid-1980s, but radios were smuggled into the prison from the early 1970s. The new electronic media created new problems of prohibition/censorship for the political captive community, which was trying hard to control the flow of information into and out of its borders. Abdallah recounts the new problematic from his critical stance toward the practice of censorship:

> . . . then in 1986 came TV . . . and the censorship worked even more than with the books, you know the TV and its scenes, although the prison authorities restricted it to Israel [Israeli Channel One], and at that time the Israeli Channel Two started to broadcast,[19] and this channel broadcasts a lot of foreign movies and there is the possibility of scenes of nudity that are much worse than the books. And Hana Mynah's [literary] pictures are nothing [compared with the audiovisual pictures of TV]. There was a big debate in the organization leadership . . . so they used the organizational censor for this purpose, when there was a scene he would turn off the TV, and the moment it ended he would turn it on . . . This whole issue created a bad atmosphere, it didn't take seriously the people who were watching the TV, and expressed mistrust in the prisoner's ability to take responsibility for himself. And not only that, it made the prisoner's imagination go wild, especially if he was a young man—what is this scene? And what did it contain? It would drain him more than protect him . . . this issue was debated again and again. (Abdallah interview, September 22, 2001)[20]

The political captives' consumption of information through TV was constrained by two main factors. First, only the Israeli channels could be received by these TV sets. Second, the times of watching TV programs and the type of programs were severely regimented by the community's leadership and institutions. Therefore, television did not compete with the written texts. On the contrary, the written word remained dominant at least up to 1993, when the whole structure of the community changed as a result of the Oslo Accords. Most of the practices of the revolutionary pedagogical institutions were conducted through reading/writing around and through the written text. Issues of sexuality are a good example and an indication of this dynamic between the written and the visual. While the ambivalent, contradictory, and conflictive expressions of sexuality of the Palestinian political captives were sometimes tabooed, there were also attempts to resituate the sexual body textually:

> . . . in 'Sqalan I read several educational essays on masturbation. The comrades did an extensive research on the subject, and wrote about its negative and harmful effects as well as about its positive ones. How many times should one do it, and how to cope with it? You know, to do sport and the like. (Abdallah interview, September 15, 2001)

The political captives dealt with their sexuality mainly textually. By projecting the sexual practices onto the text to demarcate the forbidden and the legitimate ways of coping with it, they shifted the problematic to the text, only to redirect the behavior of the body.

By using these dynamics of dominance of the text in the political captives' community, the captives show that the text of the body, and its related body of the text, seem to be located on the thin line between the repressive and the ideological apparatus. This community inculcated a Bourdieuian habitus of social and national ideologies, mainly by (con)textualizing the body of the Palestinian captive. This statement shakes my previous argument that in the colonial prison the Palestinian community used meanings/*thaqafah* to rupture the colonial time/space. The community's overinvestment in the body of the text is, in a sense, a way of bridging the colonial circumstances of the total annexation of the text of the Palestinian body by the colonial authorities. The uneven and contradictory traces and presences of the different social formations of Palestinian society, which interplay with the processes of confinement and community building in the colonial prison, lead us to problematize the analytical distinction between ideological and repressive apparatuses. The texts, as the site of practices for the ideological apparatuses, and the bodies, as the site for the practices of the repressive apparatuses, are interwoven as relational constructions on the practical expressive levels.

In this section, I started to narrate the narrative of Abdallah by presenting the actions, practices, and rituals of the educational ideological community apparatus. These basic skills of reading, writing, and frames of interpretations are located in the communal processes that defined the community's identities and ideologies through the different organizational apparatuses. Moreover, I have tried to highlight the relational and layered nature of the political captives' educational system. On the one hand, the systematic and institutionalized practices of the political captives, in their efforts to redefine the space of the colonial jail, engendered a unique form of production of knowledge/power through overinvestment in textuality and its relevant skills and practices. On the other hand, these practices and processes, while they are part of the Palestinian national movement and hence of society at large, simultaneously reside outside of the society. This dual position indicates the complex relations of body/text, which raises serious questions about the Althusserian distinction of ideological/repressive apparatuses. This position of liminality and difference will enable us to uncover the deeper workings of the Palestinian society and culture in the colonial context, by using the political captives' textual body, and the bodily text, as the example for our study.

DISSECTING THE PRACTICES FROM THE ARRESTED SOCIAL BODY

The revolutionary pedagogy of the Palestinian captive community, with its main educational system in the prison, competed with, resisted, and at times transformed both the traditional educational system of the larger society and the colonial pedagogical system. For most of his imprisonment, Radi Jirai was a key figure in his organization, Fatah, and in the community of political captives in general. During his first long sentence, from the mid-1970s to the mid-1980s, he was the cultural officer of Fatah. After the Oslo Accords in 1993, he was appointed director of the Rehabilitation Center for Former Political Captives, which was part of the newly built bureaucratic system of the Palestinian National Authority. His rich experience of political captivity, and of political activism in the larger society at different stages of the Palestinian national movement, gave him the ability to articulate and elaborate on the revolutionary pedagogical system in comparison with the colonial and traditional ones:

> As you know the traditional system in the West Bank operated according to the Jordanian curriculum and most of the teachers got their salaries from the Jordanian government. The same situation existed in Gaza, but with the Egyptians . . . They [the Jordanians and the Egyptians] did not represent us, the Palestinians, they have their own agenda. As for us, we taught the cadres in the prison about the three circles, right from the beginning with the basic skills of reading and writing; the first and most important is the Palestinian circle and it must be represented by PLO . . . then the Arab Islamic one, which should be our allies, the last one is the world beyond the first two circles, which has a different dynamic . . . The colonial authorities in the jail and in society in general wanted the Jordanian and Egyptian systems to replace the Palestinian national identity with something neutral that eliminates Palestinians . . . So educationally in the prison we had to work all the time on many fronts. (Jirai interview, August 29, 2001)

From this description and Jirai's comparative viewpoint, it seems that these processes of separating the Palestinian national identity and ideology from the colonial authorities and from the Palestinian political practices before the occupation in 1967, are linked to the processes that Taraki (1990) describes as the history of the development of political consciousness in the Palestinian territories. Farsoun and Landis (1990) analyze these practices and institutions as the sociological infrastructure of resistance. These accounts of

Taraki and Farsoun, among others, position the political captives' communal institutions in the processes of rebuilding the Palestinian national movement through political and mass organizations. But how did the practices of revolutionary/resistance education in the colonial jail create their own space/time as part of the national movement although a distinct sub community? Or is the political captives community confined within the bounds of the political culture of the dominant ideologies in Palestinian society? Has the unique constellation of the colonial jail produced its own distinctive nationalism among the range of nationalisms in the Palestinian national movement?[21]

In order to analyze the diverse patterns of relations among the different educational ideological apparatuses, it is imperative to look at Arab society in general and Palestinian society in particular as arenas of conflictive ideological formations. Many scholars have addressed these issues, the most prominent and influential of them being Hisham Sharabi. In his book *Neopatriarchy*, he argues that Arab society is torn between the traditional structures of domination, which he terms traditional patriarchy, and processes of modernization, that interact differently and in varying degrees with the existing structures of domination. Hence, for Sharabi this is an incomplete/distorted project of modernity. Sharabi delimits modernization by stating that:

> The term "modernization," used here to denote "modern" *in a patriarchal context*, has as a central characteristic, crucial to the understanding of contemporary Arab neopatriarchy that it refers to *an indigenous phenomenon resulting from contacts with European modernity in the imperialist age*. Modernization is expressed in everyday material things—dress, food, life style; institutions—schools, theaters, parliament; and in literature, philosophy and science. (Sharabi 1988, p. 22, italic in original)

But these phenomena of the processes of modernization are foreign to, in contradiction with, and in opposition to deep structures of patriarchy. In contrasting modernity and neopatriarchy as value systems, Sharabi dichotomizes them as two incompatible systems:

> As systems of value and social organization, heteronomy is based on subordination and obedience and upholds an ethic of authority, and autonomy is based on mutual respect and justice and adheres to an ethic of freedom. (Sharabi 1988, p. 43)

The schism in Arab society in general is seen as applicable to the Palestinian one in particular (Darraj 1996; Hilal 2001; Khalidi 1997; Sabbagh 1998). Sharabi's line of argumentation would lead us to articulate the political captive community, and by extension the Palestinian national movement, as a specific manifestation of the modernization processes into which the Palestinian identities and ideologies are forced due to the contact with Zionism as a specific brand of European imperialism. Thus, according to Sharabi, the building of a pedagogical revolutionary system based on modern patterns of affiliation must be situated in the processes of modernizing the Palestinian.

The processes of change and transformation of the political captive community are not necessarily attributed to modernization by the captives. For example, in the interview with Faris Qadwrah, he pointed to the changes and transformations that the community had undergone as part of the succession of older and newer generations:

> The decade of 1980s witnessed some major events for us [the political captives] . . . what the first generations of political prisoners' leaders had laid down was not good for us anymore . . . we made a new renaissance in the community . . . according to our new generation, we who had grown up under occupation. (Qadwrah interview, February 17, 2002)

As with Abdallah, Qadwrah's arguments and information lead us to the conclusion that analyzing political captivity in the frame of the modernity/patriarchy dichotomy would lead us to a narrow understanding of the deep experience of the colonial condition. The frame of analysis as developed by Sharabi and his many predecessors and followers who believe in the theories of modernization for the so-called Third World would be no more than a missed theoretical track, if not an ideological misrecognition.[22] The contact with European imperialism, in its different historical variations, did not create a distorted Europe in Palestine, or any Arab society for that matter. It initiated processes, which could and must be located historically in a conjuncture. Due to its historical contingency, this conjuncture has its own characteristics, which are not measurable in European terms of modernity/traditionality.

The Palestinian political captives' community may be seen as a test case to understand and attempt to theorize an alternative, contingent modernity. Homi Bhabha describes these processes of cultural contingency by emphasizing the multilayered nature of the localities and the borders of the national culture.

> The "locality" of national culture is neither unified nor unitary in relation to itself, nor must it be seen simply as "other" in relation to what is outside or beyond it. The boundary is Janus-faced and the problem of inside/outside must always itself be a process of hybridity, incorporating new "people" in relation to the body politic, generating other sites of meaning, and inevitably in the political process, producing unmanned sites of political antagonism and unpredictable forces for political representation. (Bhabha 1990a, p. 4)

Abdallah, through his oral and written accounts, and with his acute historical awareness, points to the heterogeneous and incoherent social forces that make up the political captives' community. The multiple cultural and social forces did not undermine or curtail the processes of building the community of captives; on the contrary, as evident from the various written accounts and interviews, it brought about dynamic and vivid processes of communal relations.

The gender issue could be the most striking of the social divisions in constituting the community of female political captives in contrast to the male captive community. The women experienced their captivity on the basis of difference, and persistent patterns of patriarchy. Iman Ahmed, a former political captive, tells the story of the women's community of political captives, which is centered more directly on and around the conflicted gender relations and issues of the larger Palestinian society:

> The community of the women political captives was not a large one . . . but in the Intifada we started to receive a different type of captives, they were not politically active or highly committed, but there were waves of stabbing soldiers and most of the stabbers were women . . . Now I had to deal with girls who had never left their homes, and on the other hand, with grandmothers . . . I couldn't enforce the party's program on them . . . so the social issues were the primary ones, courses in reading and writing, reading short stories . . . some history . . . But the worst was to deal with the families outside . . . Some would come to the prison and tell their daughter that they would never allow her to leave the house for the market even after she was released . . . Imagine the depression of that girl . . . and if it is her turn to clean the dishes for that day . . . I tried to inculcate the group spirit . . . yes the group. . . . (Ahmed interview, February 11, 2002)

Ahmed was imprisoned in 1988. For most of her nine years in captivity, she was one of the leading members of the captive community. Her description of the community of women political captives demarcates the main direct reasons for women's political captivity during the Intifada. Moreover, she positions the different context of learning and acquiring the basic skills of reading/writing, and the emphasis on, or one may say the watching eyes of, the patriarchal social control system, as the main issues that the women prisoners had to deal with. These factors do exist in the community of male prisoners, but they interact in profoundly different patterns. The families of the men see their sons as heroes, and treat them accordingly. Their reading and writing are directed toward politicization without an intermediate stage of general education, while the intermediate stage is crucial in the women's experience.

Ahmed describes the reading/writing, history, and literature courses as a preparatory stage for politicizing, that is, being a member of an organization, the women captives. Stabbing soldiers and settlers is not the dominant military action among the men. Usually men see stabbing as a desperate/primary act of resistance. These differences between the communities of men and women show us how the different cultural and social divisions and forces interplay in the processes of building communal relations. Moreover, these different communal relations are seen as modern in the sense that they are built on affiliation to a modern type of organization, namely the political party, in contrast to the traditional form of communal relations based on "blood" relations. Political captivity is seen as part of the modern Palestinian social space, which is constituted through its relation to the Israeli colonial regime.

But this stage of the Palestinian modernity is not unique in its relation to the colonial power of Israel. It has a British version, which dates back to the era of the Palestinian Mandate (Swedenburg 1995). The location of modernity in contemporary Palestinian history is the location of its Western colonization, first by the British, and then by the Zionists. Most of the intellectuals and academics concentrate their efforts on exploring the politics of Palestinian modernity, and pay little attention to the social history of the Palestinian predicament. The disastrous outcome of these efforts, and not the only one for Palestinian society, is a highly linear political historiography articulated in European and Israeli discourses, which defines Palestine as a distorted Eurocentric construct of the Modern, which then becomes one of the mechanisms for controlling it. The colonial prison system in Palestine, like the production of political knowledge about Palestine, but more intensely so, is materially and socially a modern system. Palestinians imprisoned in, colonized by, circulated in and out of this shrinking space of modernity, reshaped and reconstructed it by relying, in part, on their social formations, which are not the "pure" modern European ones.

CONCLUSION

Many processes were involved in constituting the community's alternative modernity. In this article I point to the major ones. First, a huge number of Palestinians have been imprisoned for political reasons since 1967, almost more than a quarter of the population of the West Bank and the Gaza Strip (al Hindi 2000). Second, the colonial prison is one of the most intense sites of colonizer/colonized conflict, and as such it opens possibilities for change and regeneration on the part of the colonized (Morris 2001; Qaraqi 2001; Qasim 1986; al Qaymari 1984). Third, the specific sociohistorical stage of the era after the 1967 war brought the fall of the Arab nation-state, but it also brought the rise of the Palestinian Liberation Organization and the building of its infrastructures in the occupied territories (Ajami 1981; Kimmerling and Migdal 1993). These processes initiated gradual practices of communal relations, which gradually took the shape of a distinctive community within Palestinian society and vis-à-vis the colonizer. On the one hand, these gradually built communal relations were centered, in content and form, around liberating the colonized "I," and on the other hand, this was to be achieved by resisting the colonizer "Other." But this dialectical I-Other is an ideological (mis)recognition of the national pedagogy, as Homi Bhabha (1990b) tells us.

From the interviews and the written material collected for this research, it appears that the constitutive dynamic of the Palestinian political captives community is more complex than a simplistic I-Other variation of the east–west kind of dichotomy. It seems that this is a community with dual liminality. It is exiled, in a sense deported from Palestinian society, only to be transplanted into a foreign land/space, and not any space but a liminal one, the prison of the colonial metropolitan.

From this perception of the Palestinian political captives community as a dual liminality, let us return to our opening remarks on Althusser and Feldman, in order to further our understanding of the unique social and cultural dynamics of this community. Althusser tries to give us a theoretical frame in which to articulate how a subject is reproduced by ideology, while Feldman claims that in certain conditions the subject has the agency to act transformatively in history, and hence in ideology too, but both of them have the nation-state at the back of their minds/texts. Homi Bhabha in his article "DessimiNation" argues that the Althusserian subject is an object of national pedagogy—because the nation-state controls educational systems—but at the same time, in the many constellations of time and place that exist in any culture, there are practices and performances of the object as a subject/agent. Bhabha redefines the double narrative, arguing that:

> We then have a contested cultural territory where the people must be thought in a double-time; the people are the historical "objects" of a nationalist pedagogy, giving the discourse an authority that is based on the pregiven or constituted historical origin or event; the people are also the "subjects" of a process of signification that must erase any prior or originary presence of the nation-people to demonstrate the prodigious, living principle of the people as that continual process by which the national life is redeemed and signified as a repeating and reproductive process. (Bhabha 1990b, p. 297)

Moreover, argues Bhabha, in a state of liminality not only do the subjects perform rather than being constructed by pedagogy/ideology, but they also have a wider range of political and psychological strategies for negotiating and maneuvering. In this sense of liminality, the Palestinian political captives, in relation to the wider Palestinian national society, are objects of national pedagogy, but, and perhaps more importantly, they are the subjects who create this same nationalism by struggling to liberate the collective and the individual "I" (Harlow 1987). The same actions, practices, and rituals of reading/writing and interpretation, which interpellate the Palestinian subject as a pedagogical object, when performed in the context of political captivity, that is, liminality, become a revolutionary, educational, ideological, and communal apparatus. Moreover, as positioned simultaneously by the national "I" and by the colonizer "Other," at the edges of the colony–metropolitan continuum, the Palestinian political captives' writing/reading transformative agency (Hafez 1993) decentralizes the authority of the hegemonic metropolitan and the homogenized colony.

In an interview with one of the first generation of leading political captives, Rami Abid told me a story in which both the homogenized practices of masculinity of the Palestinian society and the hegemonic practices of confinement were simultaneously superseded. Evidently, in the late 1960s he already had the sense of a revolutionary who destabilizes both the metropolitan and the colony.

> I was brought to the military court in Ramallah . . . to testify . . . they put me in the dungeon with other captives . . . One of them had a book that I was dreaming of reading . . . you know dreaming . . . I copied it by hand, I didn't sleep for three nights copying it . . . then I arranged it in capsules . . . Some

moments for you as a man [in terms of masculine sexuality] you are not ashamed to do things that are usually shameful . . . but I didn't think twice, I had the five capsules so I put them in my ass . . . It wasn't easy . . . everybody was watching, nobody joked or anything like that, people were basically shocked, but later I became a model for them. . . . (Abid interview, October 2, 2001)

The story of Rami Abid highlights the main arguments of this chapter. The unique relations of text and body that developed in the social space of the colonial jail collide with the accepted boundaries set by the colonial authorities as well as by Palestinian society. The body of the political captive is read differently. The body becomes an open text for new interpretations, defying its confinement to the one interpretation imposed by the colonizer. However, this act of defiance requires new skills of reading/writing and interpretation. In this manner the text comes to be the dominant social space of resistance in the colonial jail, which basically aims to annex the body, the text, and their material context. This dynamic of resistance through rereading and rewriting the text of the body and the body of the text, although intensified in the colonial jail, is not confined to it in the modern Palestinian national movement.

The political captive and the armed struggler are seen as heroes in the Palestinian national discourse. According to this discourse, a hero is a Palestinian who stops his ordinary life only to be in a constant liminal space/time of resisting the colonizers. Moreover, the hero will succeed, in his mythical time/space alterity, by the constant retelling of his actions and deeds in relation to the "Other," not by himself, but by the Palestinians as a collective (Harlow 1996). This structure/trope of national heroism was/is reproduced and produced at the same time in the two liminal Palestinian national spaces, the armed struggle and political captivity. Each domain has its own revolutionary pedagogy, namely its own educational organizational apparatus (Althusser 1971). The split, as Bhabha calls it, is expressed in the fact that at the moment of his creation by the Palestinian subject, the Palestinian hero becomes a national pedagogical object.

The moment of constituting a hero by a specific discourse is the moment of disseminating him throughout the larger society as a communicative event. The elevation of certain sets of practices—in our case political captivity and the armed struggle—into heroism is linked to the interrelation of text and body. The rituals that transform the accepted divisions between the body and the text in the mundane and ordinary time generate the dynamics of heroization, culminating in a communicative act in order to close the cycle on the social and cultural levels. Ata al Qaymari vividly puts this cycle into a narrative:

The first year and half they put me with the criminal prisoners because I was young . . . I couldn't wait to be transferred to see Rami and Mahmud, they were mythic figures for me . . . In 1973 I met Rami, Mahmud wasn't there, he was already out . . . he was my superior, he is a real struggler, he was my hero . . . Later, I understood their mistakes . . . then I myself was seen as a hero by the newcomers, they were reading my articles and hearing about me . . . the wonder child who did his first military operation at 14 years old with no organizational help. (al Qaymari interview, January 18, 2002)

This chapter has illuminated the ways in which culture works in the historical moment of building the ideologies and identities of a nation-state. Using the example of the educational apparatuses of the community of Palestinian political captives, this chapter reopens a passage of local and specific history in the Palestinian national movement in order to explore the inner mechanisms of rebuilding the national ideologies and identities. Basically, the junction of the colonial jail is redefined to examine the Palestinian body and the Palestinian text. The overinvestment in skills becomes comprehensible through the reinterpretation of the social boundaries, spaces, and temporalities of "Who is the Palestinian?" Setting heroes is a pedagogical act aimed at mass mobilization and control, which then is seen by the national masses as a historical agent in the national theater of consciousness (Pecheux 1994).

These experiences of revolutionary pedagogy in the prison had long lasting implications for Palestinian society. To understand these implications, it is necessary to research the different social sites of Palestinian society in relation to the experience of captivity. Prominent, in this regard, is the transformational processes that the Palestinian national movement, and the PLO in particular, went through since the early 1990s. Most of the researchers of the Palestinian society and politics argue that the PLO is almost a dismantled organization, and claim that the national movement lost its grip in the Palestinian society (Farsoun and Zacharia 1997). These arguments notwithstanding, looking at the political captivity issues in the era after Oslo, one could see that the Palestinian National Authority is, at least, ambivalent toward them. On the one hand, it incorporated the leadership of the political captives into its institutions, and created different formal bodies to deal with the former political captives, such as the ministry of political captives, centers for rehabilitations, and social clubs. On the other hand, and contrary to the expectations of the different Palestinian social and political groups, the PLO's negotiation teams did not put political captivity as a critical item in the agenda of negotiation with the Israelis. One of the claims was that the leadership of the PLO did not realize the importance of the issue of captivity for the Palestinian masses in the West Bank and the Gaza Strip. These ambivalent attitudes, and shifting practices could indicate the dire situation of the Palestinian national movement in the late 1990s. Certainly enough, though, it did not disappear but is taking a different shape as the last Intifada, which erupted in September 2000, amply shows. Finally, it remains to be seen how other national liberation movements, which were/are linked directly or indirectly to the Palestinian one, rebuild their own national identities and ideologies interactively with each other's experiences of political captivity. Notable among these are the Irish and the South African experiences of political captivity as many researchers indicate a similar primacy of knowledge production in building their communities inside the prisons (Beresford 1987; Buntman 1998).

NOTES

1. There are different terms in Arabic to describe the prison and the jail. In this article, I use the term political captivity because it is used by the Palestinians to describe the phenomenon of imprisonment by the Israelis and its related sociohistorical processes. The term in Arabic is *al asr al syasy*. Although the captive as a specific individual is *mutaqal*, which means literally "detained," this is used only for political captives in the Palestinian context.

2. Arabic names and words are transliterated here according to the Library of Congress rules except for names that have already been published in a different form in English, such as Intifada.

3. The Palestinian Intifada of 1987 had many determining effects on the Palestinian national movement. One of the main processes it set in motion was the shifting of the main arena of struggle for liberation to the West Bank and the Gaza Strip. Part of this process was the rise of Palestinian intellectuals living in Palestine to the upper ranks of the Palestinian leadership. Lisa Taraki, who works at Bir Zeit University, is one of this intellectual group. In addition, the relations between the intellectual political captives and these groups of academics and political activists expanded and deepened during the Intifada of 1987.

4. For historiography, see al Qaymari (1984), Qasim (1986), and Abu il Haj (1992); for literary criticism see Abdallah (1994), and al Jwhar (1997), among others; for political essays, see Qaraqi (2001). In the Palestinian press the issues related to political imprisonment are discussed almost daily.

5. This chapter does not address the Islamic movements because they came in at a later stage, and because of the different dynamics and structures of their communal relations.

6. The number of prisons changed as new ones were opened and others closed, but there are usually about 20 active prisons. Moreover, in spite of the confinement and the severe regimentation, the Palestinian political captives managed to develop sophisticated networks of verbal and written communication with the outside world, and between the different colonial prisons. The development of communication networks was a crucial step in the process of building this community.

7. One of the recurring arguments in the many interviews that I conducted with former political captives from the different Palestinian organizations is that one should die rather than let any piece of paper or pen be taken from him by any of the representatives of the prison authorities. These attitudes and values should be understood in the context of resistance and liberation. Liberating a paper in the colonial jail has a social value that echoes far beyond the paper as such.

8. Usually the leadership of the political captives community resides in one of these prisons, such as 'Sqalan or, later, Jnayd prison in Nablus. The Israel prison authorities have made several attempts to disperse the leadership of the community, most notoriously when the Israeli administration of the prison system opened Nafhah prison in 1980 and transferred to it some 80 political captives whom they regarded as the "ringleaders" of this community.

9. After the occupation of the West Bank and the Gaza Strip in 1967, the educational system continued to follow the Jordanian curricula in the West Bank and the Egyptian one in the Gaza Strip. An educational officer in the Israeli colonial administration amended and changed the texts and the teachers according to the colonial administration practices.

10. Hasan Abdallah's publications include collections of short stories, such as *A Bride and a Groom in the Snow* (1993), documentary works and commentaries on prison literature, including *The Literary Production of Political Imprisonment* (1994), examples of prison journalism collected in *A Journalism that Challenged the Chain* (1996), and many other articles and interviews in the local newspapers.

11. The language of the interview was Arabic. Hasan used vernacular and standard Arabic in the interview. Moreover, he used different vernacular dialects. In the translations for this article, I have tried to convey the atmosphere and the dynamic of each variety of Arabic wherever possible.

12. The calendar of the political captives contains many memorial dates, which the captives usually celebrate with festivities and speeches. The prison authorities regularly try to prevent these celebrations.

13. Each prison has a public library. Usually, each Palestinian organization in the prison has its own library in addition to the public one. According to different testimonies, the public library of the Jnayd prison had around 8,000 titles, most of which had been brought in by the prisoners.

14. Hana Mynah is a famous and prolific Syrian novelist.

15. Many of the political captives who are known as experts are also leading figures in the community. Moreover, they took leading positions in the Palestinian National Authority or the private sector. For example, Faris Qadwrah is an elected member of the Palestinian legislative council; Jibryl Rajwb was the head of Preventive Security.

16. As mentioned earlier, this article explores the Palestinian secular national movement, and does not address the Islamic movements. Needless to say, the latter have their own perspectives, which at times merge with the national ones and at others profoundly contradict them.

17. See Balibar and Wallerstein (1991) for theoretical discussion and Doumani (1995) for discussion on the Palestinian social history.

18. An Israeli magazine, established by Uri Avneri, and known for its leftist outlook.

19. Until the late 1980s the Israeli Channel One, which is owned and run by the state of Israel, was the only legal TV channel in Israel. Channel Two was the first private TV channel in Israel.

20. In the early 1990s the national organizations stopped censoring TV programs. But the Palestinian Islamic movements, which I do not deal with in this article, still practice censorship of TV programs.

21. See Darraj (1996) for a brilliant critique of the dynamics of culture and politics in the Palestinian national institutions.

22. Hisham Sharabi and many other Arab intellectuals are eager to see Arab societies in general, and the Palestinian one in particular, "develop" into modern ones, "modern" being the neoliberal modern. It seems that his ideological projections "distort" his analytical intervention, and not that the Arabs are a distorted modern project as such.

References

Interviews

Ahmad, Iman, Bir Zeit, February 11, 2002.
Abdallah, Hasan, Ramallah, September 6, 2001; September 15, 2001; September 22, 2001.

Abid, Rami, Ramallah, October 2, 2001.

Jirai, Radi, Ramallah, August 29, 2001.

'Iayan, Muhammad, Jerusalem, December 6, 2001.

Qadwrah, Faris, Ramallah, February 17, 2002.

al Qaymari, Ata, Jerusalem, January 13, 2002; January 18, 2002.

Publications

Abdallah, H. 1993. *A Bride and a Groom in the Snow*. Ramallah: Alqalam House (Arabic).

———. 1994. *The Literary Production of Political Imprisonment: A Historical Analytical Study*. Jerusalem: Alzahra' Center for Research (Arabic).

———. 1996. *A Journalism that Challenged the Chain*. Ramallah: Al Mashriq Center for Research (Arabic).

Abu il Haj, F. 1992. *The Knights of the Intifada Talking from Behind the Bars*. Jerusalem: Jamiiat Al Dirast Al Arabieh (Arabic).

Ajami, F. 1981. *The Arab Predicament: Arab Political Thought and Practice Since 1967*. Cambridge: Cambridge University Press.

Althusser, L. 1971. *Lenin and Philosophy, and Other Essays*. New York: Monthly Review Press.

Amnesty International. 1990. *Report*.

Balibar, E. and Wallerstein, I. 1991. *Race, Nation, Class: Ambiguous Identities*. London: Verso.

Beresford, D. 1987. *Ten Men Dead: The Story of the 1981 Irish Hunger Strike*. New York: Atlantic Monthly Press.

Bhabha, H. 1990a. "Introduction: Narrating the Nation." In *Nation and Narration*. Bhabha, ed. London: Routledge.

———. 1990b. "DissemiNation: Time, Narrative, and the Margins of the Modern Nation." In *Nation and Narration*. London: Routledge.

Buntman, F. L. 1998. "Categorical and Strategic Resistance and the Making of Political Prisoner Identity in Apartheid's Robben Island Prison." *Social Identities* vol. 4, no. 3, pp. 417–441.

Darraj, F. 1996. *The Poverty of Culture in the Palestinian Institution*. Beirut: Dar Al Adab (Arabic).

Doumani, B. 1995. *Rediscovering Palestine: The Merchants and Peasants of Jabal Nablus, 1700–1900*. Berkeley: University of California Press.

al Dyk, A. 1993. *The Intifada Society*. Beirut: Dar Al Adab (Arabic).

Farsoun, S. K. and J. M. Landis. 1990. "The Sociology of an Uprising: The Roots of the Intifada." In *Intifada: Palestine at the Crossroads*. J. R. Nassar and R. Heacock, eds. New York: Praeger.

Farsoun, S. K. and C. E. Zacharia. 1997. *Palestine and the Palestinians*. Boulder, Colo.: Westview.

Feldman, A. 1991. *Formations of Violence: The Narrative of the Body and Political Terror in Northern Ireland*. Chicago: The University of Chicago Press.

Foucault, M. 1978. *Discipline and Punish: The Birth of the Prison*. New York: Vintage Books.

Hafez, S. 1993. *The Genesis of Arabic Narrative Discourse: A Study in the Sociology of Modern Arabic Literature*. London: Saqi Books.

Harlow, B. 1987. *Resistance Literature*. New York: Methuen.

————. 1996. *After Lives: Legacies of Revolutionary Writing*. London: Verso.

Hilal, J. 2001. *The Formation of the Palestinian Elite: From the Rise of the Palestinian National Movement to the Constitution of the National Authority*. Ramallah: Muwatin Press (Arabic).

Hiltermann, J. 1991. *Behind the Intifada: Labor and Women's Movements in the Occupied Territories*. Princeton: Princeton University Press.

al Hindi, K. 2000. *The Democratic Practice of the Palestinian Political Prisoners Movement*. Ramallah: Muwatin Press (Arabic).

al Jwhar, Z. 1997. *The Poetry of Prison in Palestine, 1967–1993*. Ramallah: The House of Poetry (Arabic).

Khalidi, R. 1997. *Palestinian Identity: The Construction of Modern National Consciousness*. New York: Columbia University Press.

Khalyl, M. L. 1989. *The Imprisonment Experience in the Israeli Prisons*. Amman: Ibin Rushd Press (Arabic).

Kimmerling, B. and J. Migdal. 1993. *Palestinians: The Making of a People*. Cambridge, Mass.: Harvard University Press.

Morris, B. 2001. *Righteous Victims: A History of the Zionist-Arab Conflict, 1881–2001*. New York: Vintage Books.

Pecheux, M. 1994. "The Mechanism of Ideological (Mis)recognition." In *Mapping Ideology*. S. Zizek, ed. London: Verso.

Qaraqi, I. 2001. *The Palestinian Political Prisoners in the Israeli Prisons after Oslo: 1993–1999*. Birzeit: Birzeit University Press (Arabic).

Qasim, A. 1986. *The Captivity Experience in the Zionist Jails*. Bierut: Al Amal Press.

al Qaymari, A. 1984. *The Prison is Not for Us*. Jerusalem: n.p. (Arabic).

Sabbagh, S., ed. 1998. *Palestinian Women of Gaza and the West Bank*. Indianapolis: Indiana University Press.

Saleh, S. A. 1990. "The Effects of the Israeli Occupation on the Economy of the West Bank and the Gaza Strip." In *Intifada*. Nassar and Heacock, eds. New York: Praeger.

Sharabi, H. 1988. *Neopatriarchy: A Theory of Distorted Change in Arab Society*. Oxford: Oxford University Press.

Swedenburg, T. 1995. *Memories of Revolt: The 1936–1939 Rebellion and the Palestinian National Past*. Minneapolis: University of Minnesota Press.

Taraki, L. 1990. "The Development of Political Consciousness among Palestinians in the Occupied Territories, 1967–1987." In *Intifada* Nassar and Heacock, eds. New York: Praeger.

8

Teaching to Hate

The Hindu Right's Pedagogical Program

Nandini Sundar

In 1992, "Hindu" *kar sevaks* or "volunteers for god" under the leadership of the Bharatiya Janata Party (BJP) and other fronts of the Hindu chauvinist organization, the Rashtriya Swayamsevak Sangh (RSS),[1] demolished the Babri Masjid, a fifteenth-century mosque in Ayodhya, sparking off a round of violence across the country. They claimed the mosque had been built over an earlier temple commemorating the birthplace of the Hindu god, Ram, a claim that secular professional historians have contested.[2] For many children in RSS run schools, however, this divisive, painful, and bloody propaganda has long had the status of "facts," which they learn as part of "history" or "general knowledge." For instance, the *Sanskriti Gyan Pariksha Prashnotri* or a primer for a "Cultural Knowledge Exam," which all students in RSS schools take and for which they get a certificate, has the following list of questions and answers:

Q: Which Mughal invader destroyed the Ram temple in 1582?
A: Babur
Q: From 1582 till 1992, how many devotees of Ram sacrificed their lives to liberate the temple?
A: 350,000.
Q: When did the program of collecting bricks for the Ram Mandir begin?
A: September 30, 1989.
Q: When did the Karsevaks fly the saffron flag on Ramjanmabhoomi?
A: October 30, 1990.

Ten years after this demolition, with the BJP now in power as the leading partner in a coalition government, a key arena of struggle within India is the extent to which RSS educational philosophy is allowed to become official policy.[3] If schools are one of the modes by which nations imagine and reproduce themselves, debates over schooling systems—availability, cost, curriculum, language, pedagogical techniques—are, at heart, debates over the style and content of this imagining. This chapter is an attempt to examine the ways in which Hindu Right notions of citizenship, nationhood, and patriotism are shaping the national debate over education in India and how these notions are transmitted through the RSS's own widespread and growing network of schools.

With the BJP in power, the RSS has been able to position its own people in national educational bodies, such as the National Council of Educational Research and Training (NCERT) and the University Grants Commission (UGC)[4] and thus determine the official educational agenda for the country. Much before it came to power, however, the RSS was cognizant of the centrality of education to any project of gaining power and reorienting the political arena. Its educational front, Vidya Bharati, runs one of the largest private network of schools across the country. As of March 2002, Vidya Bharati had 17,396 schools across the country (both rural and urban), 2.2 million students, over 93,000 teachers, 15 teacher training colleges, 12 degree colleges and 7 vocational and training institutions.[5] Although Vidya Bharati schools follow state board exams and state prescribed textbooks, they have additional subjects and cocurricular activities, during which the bulk of "indoctrination" takes place. In addition to Vidya Bharati, which caters mostly to a lower-middle-class and upper-caste base, other RSS fronts providing education include the Vanvasi Kalyan Ashram (VKA), which specializes in welfare schemes for *adivasis* (lit. original inhabitant),[6] including hostels for school children; Sewa Bharati, which works among *dalits* (scheduled castes);[7] and the Ekal Vidyalaya Foundation, which runs single teacher three hour centers for preschool children where they are taught the rudiments of reading and writing, Sanskrit and *sanskars* (good behavior).

When it comes to debating the nation's past or understanding contemporary society at the official level, the BJP shares with other conservative governments certain preferences and omissions in deciding curriculum content: for them, social science, especially history, is intimately connected with inculcating national identity and patriotism (Hein and Selden 1998; Kumar 2001; Nash et al. 2000; Nelson 2002). Most often, moreover, the "nation" is identified with dominant groups in society and only narratives that value the role of these groups positively are seen as patriotic.[8] Patriotism, in this version, also means extolling the virtues of one's own country over that of others and glossing over its negative phases (Nash et al. 2000). In the United States, this involves describing America as a consistent defender of freedom and democracy all over the world and minimizing discussions of slavery or the extermination of native Americans (Hellinger and Brooks 1991), in Japan it involves glossing over wartime atrocities (Nelson 2002), and in India it takes the form of glorifying an ancient Hindu past and minimizing caste oppression. Ideological conflicts over the shape and content of history or civics textbooks inevitably involve debates over the emphasis given to certain people—dominant groups versus women and minorities—or certain periods in history over others.

In terms of the pedagogy transmitted by the RSS's own schools, however, the more relevant model for comparison appears to be fascist schooling, although there are significant differences in that the RSS's control over the state is still far from complete, and is further complicated by the country's federal structure and the creation/reinforcement of distinct religious and ethnic populations through separate schooling systems. Under fascism, while the curricular content is geared to define nationhood in terms of particular attributes, the major work of cultural demarcation is done through extracurricular activities and the everyday rituals that punctuate the school day such as assembly, attendance, uniforms, and so on (Blackburn 1985; Kandel 1935; Mann 1938; Minio-Paluello 1946). As against the mere silencing of the diverse cultures and contributions of women, minorities, or workers found in conservative visions of education, the fascist school curriculum involves an active manipulation of historical evidence in order to foster hatred for and violence against minorities. While denigrating people of other countries is common enough in all school curricula, especially countries with which one has been at war, fascist schooling is also marked by the construction of enemies within. Finally, while fascist schools do not have a monopoly on militarist education,[9] preparing children for national "defence" is intrinsic to the fascist project (Giles 1992; Wolff 1992).

The first section of this chapter describes the RSS's attempts to appropriate the national and official educational arena by introducing communal, and often simply false, interpretations of history and society. The second section, based on ethnographic fieldwork in Chhattisgarh state in central India, shows how apart from just being overtly "Hindu," RSS education aims to produce students geared to thinking of non-Hindus as enemies. Such education "succeeds" in attracting ordinary non-RSS Hindu students, however, because of the failures of the state educational system.

CONTROLLING NATIONAL
EDUCATION: THE TEXTBOOKS WAR

In India, education has rarely occupied much space in the public sphere. Within this realm, the focus has mostly been on issues of literacy and access to schooling rather than curricular content.[10] Indian government schools are the bedrock of the educational system, especially in rural areas, although the number of private schools is fast expanding. However, both government and private schools generally follow government-mandated curricula, and use government produced textbooks. In India's federal structure, in addition to the national school boards such as the Central Board for Secondary Education, individual states have their own school boards that conduct certification exams. States also bring out their own curriculum and textbooks, but these are usually modeled on the national curriculum framework and textbooks produced by the National Council for Education and Training (NCERT), which is technically an autonomous body, but in practice, part of the government. The federal government is thus a key player in determining educational content and performance across the country. While the quality of textbooks has been pedagogically patchy in the past (Kumar 2001) and the extent to which schools have been really secular can be debated,[11] in principle, at least, the federal government was committed to an educational policy that reflected the constitutional definition of India as a secular state.

The appointment of an RSS ideologue, Murli Manohar Joshi, as Minister of Human Resource Development (the portfolio that includes education) has introduced a qualitatively new dimension to the educational scene. In 2000–01, the NCERT issued a National Curriculum Framework for School Education, which according to critics, reproduced the RSS's ideological agenda (Delhi Historians' Group 2001). The essence of the new educational philosophy is summed up in the terms "Indianize, nationalize and spiritualize."[12] For the RSS, even as schooling is seen as an overt and directed instrument of political and cultural transformation aimed at creating a unified "Hindu" nation, its educational philosophy is far from revolutionary,[13] and seeks to overturn even the limited advances toward greater caste, class, and gender equality since independence. Much of it is aimed at reinforcing the existing social order in the face of growing challenges from below—from women, lower classes, and lower castes—relying on an appeal to a glorious past, before it was destroyed by "Muslim invaders," "Christian missionaries," and "Westernization."

The controversial National Curriculum Framework for School Education, drawn up by the NCERT in 2000, bears quoting in detail:

> Traditionally, India has been perceived as a source of fulfillment—material, sensuous and spiritual, consisting primarily of an agrarian society, the social design of which emphasized self-sufficiency, contentment and operational autonomy for each village . . . The social matrix was congruent with the economic design based on the principle of distributive authority given to each village unit . . . A sizeable segment of the contemporary Indian society seems to have distanced itself from the religio-philosophic ethos, the awareness of the social design and the understanding of the heritage of the past . . . the structure of the authority of the Indian agrarian society has been disturbed. *An individual in the formal work system could exercise authority over those who were otherwise his superiors in age and in the societal structure.* In the agrarian society, successive generations followed the occupation as well as the goal sets of the family or caste at large. However, technological developments later introduced new professions and consequently new goal sets emerged. *In contrast to the joint and extended family system, the society now is witnessing the phenomenon of nuclear families, single parents, unmarried relationships and so on.* (NCERT 2000, pp. 3–4; emphasis added)

Despite the formulaic appeal in the very next paragraph to the "national goals of secularism, democracy, equality, liberty, fraternity, justice, national integration and patriotism," it is impossible not to note the sense of loss and dismay that greets the changes in this idealized agrarian society, such as the loss of caste authority when a low caste person is in a position of bureaucratic power or when women are no longer bound by the patriarchal joint family.

A petition by three activists/citizens, pleading that the NCERT had not followed the correct procedures for consultation with the states before preparing the curriculum and that it sought to introduce religious teaching was, however, rejected by the Supreme Court. Article 28 of the Constitution that states "no religious instruction shall be provided in any educational institution wholly maintained out of state funds" was interpreted by the Judge to allow for teaching *about* religions. In addition, Justice Shah expounded on the virtues of religion as the source of ethics (values) in a society fast losing its social moorings: "for controlling wild animal instinct in human beings and for having civilized cultural society, it appears that religions have come into existence. Religion is the foundation for value base survival of human beings in a civilized society".[14]

Having won its case, the NCERT released its history and social science textbooks. Although textbooks in other subjects have also been criticized (Menon 2003) they have not earned the same kind of attention as history texts—in part because of the centrality of history to the RSS's own project of rule. In a logic where the past is seen as determining current identity, rights to citizenship, and raising questions of historic reparation and justice, the teaching of history is inevitably deeply political (Kumar 2001; Nash et al. 2000). What Hitler laid down for the teaching of history in *Mein Kampf* might well apply to the RSS agenda today: "History is not studied in order to know what has been, but one studies history in order to find in it a guide for the future and for the continued preservation of the nation itself" (quoted in Kandel 1935, p. 65). In part, the greater focus on history is also due to the strength and sophistication of contemporary historical scholarship in India, and its success in challenging colonial and communal narratives (Lal 2003).

The NCERT social science/history textbooks are not only shockingly low on both grammar and fact, but also reflect many of the RSS's pet themes, particularly the urge to prove that Indian civilization is synonymous with Hinduism, which in turn is synonymous with the "Vedic civilization."[15] This Vedic civilization is thought to be the font of all things great in the world including the discovery of the zero, while all the evils that beset India can be traced to foreigners, including Muslim invaders and Christian missionaries. In the textbook on medieval India, the exactions of the Sultanate Rulers or the Mughals are exaggerated and portrayed in anti-Hindu terms, and their contributions to society, culture, and polity are ignored. The idea that the Babri masjid was built on an earlier temple is given textual sanctity: "the sites (for Babur's mosques) were carefully selected . . . Ayodhya was revered as the birthplace of Rama" (Jain 2002, p. 134). It is not difficult to see the parallels with the version of history coined by the RSS's late chief and ideologue M. S. Golwalkar, in his *We or Our Nationhood Defined*: "Ever since that evil day when Moslems first landed in Hindusthan, right up to the present moment the Hindu nation has been gallantly fighting on to shake off the despoilers" (Golwalkar 1939, p. 12).

In his remarkable book, *Prejudice and Pride*, Krishna Kumar compares Indian and Pakistani textbooks on their narrative of the freedom struggle, and finds both lacking in important ways. In projecting the freedom struggle as a secular progression, rudely interrupted by partition, and by focusing mainly on political events and personalities, the earlier NCERT textbook on modern India, Kumar argues, did not enable children to understand the processes and sociological factors that led to communalization and partition. Nor do they get a sense of why a secular constitution meant so much for India's leaders (Kumar 2001). In the new NCERT textbook for class IX, however, there is even less attempt to understand ideals and processes. Instead, the blame is clearly assigned. *Contemporary India* spends considerable time on the role of the Muslim league in causing partition, and the perfidy of the communists for supporting the Allies in World War II, while omitting any mention of the RSS or the Hindu Mahasabha's contribution to communalism and partition. There is no word on Gandhi's assassination by RSS sympathizer, Nathuram Godse. With such remarkable passages as "The task of the framers of the constitution was very difficult. Their foremost job was to ensure the integrity of the country taking into account the presence of Pakistan within India herself," children are easily led to see Muslims as fifth columnists and not fully Indian. Much of this is again presaged in Golwalkar:

> In this country, Hindusthan, the Hindu race with its Hindu Religion, Hindu Culture and Hindu Language (the natural family of Sanskrit and her offsprings) complete the Nation concept; that, in fine, in Hindusthan exists and must needs exist the ancient Hindu nation and nought else but the Hindu nation. All those not belonging to the national, i.e. Hindu Race, Religion, Culture and Language, naturally fall out of the pale of real "National" life . . . . Only those movements are truly "National" as aim at re-building, revitalizing and emancipating from its present stupor, the Hindu nation . . . . All others are either traitors or enemies to the National cause . . . . (Golwalkar 1939, pp. 43–44)

Contemporary India covers the scope of world history from the fifteenth to the twenty-first centuries in ten pages, presumably on the assumption that 14-year-olds cannot cope with more. The Russian revolution is dismissed in a couple of lines: "Many generation-old rule of the family of Czars was swept away by a coup led by Lenin, the leader of the Bolshevik Party. This political change was presented to the world as an ideological revolution rooted in Marxism and Communism" (NCERT 2000, p. 9). As for Hitler, we have this remarkable paragraph:

> The German nationalism which had developed a superiority complex about the purity and antiquity of its so-called Aryan blood was smarting under the humiliating terms imposed on it under the Treaty of Versailles. The German frustration gave birth to the personality of Adolf Hitler who created the Nazi party. The ideology of the Nazi party was a sort of fusion of German nationalism and socialism. The rising tide of German nationalism was seething with an ardent desire of revenge. The Germans readily accepted Hitler as their leader and surrendered to his dictatorship. (NCERT 2000, p. 10)

There is no mention of the Holocaust, and perhaps this is not surprising in the light of what Golwalkar had to say on the subject:

> To keep up the purity of the Race and its culture, Germany shocked the world by her purging the country of the Semitic Races—the Jews. Race pride at its highest has been manifested here. Germany has also shown how well nigh impossible it is for Races and cultures, having differences going to the root, to be assimilated into one united whole, a good lesson for us in Hindusthan to learn and profit by. (Golwalkar 1939, p. 35)

When challenged by historians on his seeming admiration for Hitler, Hari Om, the author of *Contemporary India*, argued that their criticism stemmed from a desire to cover up the far worse crimes of communism: "To the Indian Leftists, the Fascists were some kind of red herring. By concentrating on the crimes of Hitler and Mussolini, they hoped to divert attention from their own misdeeds which, quantitatively as well as qualitatively, were of bigger proportions" (Lal et al. 2003, p. 231).

STATE FAILURE IN SCHOOLING: BREEDING GROUND FOR SECTARIANISM

While it is important to understand the Hindu Right agenda on education and explore how far this schooling actually works to produce Hindu chauvinist identities, it is equally important to see this as a product of the wider context of state schooling. Despite Article 45 of the Directive Principles of the Constitution urging all states to provide "free and compulsory education for all children till they reach the age of fourteen years," and a recent constitutional amendment upgrading education to a fundamental right, literacy rates in India are comparatively low (65.4 percent in the 2001 census). In any case, even in the face of reassuring statistics on the number of schools opened, teachers hired or children enrolled, the micro level reality remains grim. In 1999, the Public Report on Basic Education (PROBE) in the low literacy states of central India found widespread teacher absenteeism, leaking roofs, nonexistent toilets, no drinking water, no blackboards, and no educational materials such as textbooks and maps. The report wryly notes "the only available teaching aid available in all schools is a stick to beat the children" (PROBE 1999, p. 42). In addition to these problems are those peculiar to areas inhabited by adivasis, such as the language gap between students and teachers who do not speak any of the local languages, blatant discrimination or at the very least unequal treatment by teachers compared with non-adivasis or upper caste students, and general condescension that makes the educational experience particularly alienating for most tribal children (Nambissan 2000; Nanda 1994; PROBE 1999).

Teachers, especially in adivasi areas, often blame parents for not sending their children to school, either because they are too "backward" to realize the importance of schooling or because they need to send their children to work.

The same report as quoted above also states, however, that parental motivation to send their children to school is high across the country. In fact it is not the opportunity cost of child labor foregone that is the problem but the cost of sending children to school, as well as the poor quality of education and the sense of inferiority generated in tribal children that makes it not worth the expense (Furer-Haimendorf 1982; PROBE 1999).[16]

In other words, even among those populations historically alienated by the schooling system, there is a relatively strong and widespread desire for education as a means to challenge existing social inequities and to lay claim to the attributes of citizenship that the state promises, despite their experiences of the state in practice. One consequence of this widespread desire for education and the lack of matching state initiative has been the considerable increase in private schooling, both of the religious nonprofit and the supposedly secular profit-making variety. In many places, particularly urban or semiurban areas, this has exacerbated social differentiation, with the poor being confined to vernacular government schools and anyone with the slightest ability to pay sending their children to private schools, preferably "English medium" ones. Increasing class and communal divisions, promoted through differential schooling, thus diminish the promise of a more meaningful common citizenship held out by higher literacy levels (Jeffery et al. 2002; Vasavi 2000).

Multiple Systems: RSS, Catholic and State Schools in Jashpur District, Chhattisgarh

In order to appreciate the context in which RSS schools have taken root and the message they impart, it is instructive to look at the actual functioning of such schools. Textbooks are only a small part of pedagogical practice, and their reception cannot be understood without looking at the specific setting in which they are taught. As Luykx shows us: "educational processes . . . are fundamentally, cultural processes" (Luykx 1999, p. xxxiii). This section, accordingly, aims at fleshing out this notion of pedagogy as cultural interpellation, based on fieldwork conducted in Chhattisgarh state primarily, but not only, in Kunkuri, the headquarters of a development block in Jashpur district. Like other small urban concentrations (really glorified villages) in a predominantly adivasi belt, Kunkuri has a significant number of traders, mostly from the Marwari and Jain communities but also some Muslim families. A number of government staff also live here. The villages nearby consist of Uraon adivasi cultivators most of whom are now Christians. Unlike most similar localities in this state, Kunkuri has a proliferation of schools and is known as something of an educational center in the district. This is primarily because it is home to one of the oldest and best Jesuit boarding schools in Chhattisgarh, the Loyola boy's high school.

In the late 1940s, the Vanvasi Kalyan Ashram (VKA) was founded here to counter Christian proselytization, setting in motion a process of competitive schooling. In recent years the Jesuits have also started a coeducational English medium day school in the Loyola premises. A Catholic girls high school,

Nirmala, run by the Order of St. Anne's, has a small boarding house attached to it. Kunkuri has two government high schools with hostels (one for girls and one for boys), as well as a government middle school. It also has a Saraswati Shishu Mandir (run by Vidya Bharati), which is at present only from primary to middle school but has plans to expand to high school.[17] The Saraswati Shishu Mandir has no hostel, but some of the students stay in the Vanvasi Kalyan Ashram hostel. Some VKA hostellers go to Loyola as well.

The schools are distinctive not just in terms of their extracurricular content, but the social composition of both students and teachers (see tables 8.1–8.3). In both the Catholic and RSS case, there is some degree of boundary policing for their own communities. The Catholic schools cater primarily to Christian adivasi students, who are mostly the children of both peasants and salaried government employees. Large-scale recruitment to the army and to the Assam tea gardens, from World War II onward, encouraged by the Church, brought greater prosperity and also increased income differentiation among the local adivasis. Thanks to the education they get from these schools, many of the Uraons in this area have been able to take advantage of reservations for adivasis in government jobs and Loyola graduates are estimated to occupy about 20–40 percent of the state administration posts in Raipur, the state headquarters. In addition to the priests who teach, almost all the teachers are Catholic. Many of the business families have also traditionally sent their children to Loyola, because it was the best education on offer in the area. However, now that there are alternatives in the form of the government school and Saraswati Shishu Mandir, which also have good results, many parents have started sending their children there.

Table 8.1 Social composition of students of Loyola, Saraswati Shishu Mandir (SSM), and Government Schools (2001–02)

	Scheduled tribes	Scheduled castes	Other backward castes	General	Total
Loyola (6–12)	1,042	37	162	42	1,283
Loyola Hostel (6–12)	422	—	47	4	473
Loyola English Medium (1–5)	NA	NA	NA	NA	270
Loyola English Medium (6–9)	NA	NA	NA	NA	124
SSM Primary (1–5)	100	50	140	105	395
SSM (Middle and High 6–10)	50	30	316	10	406
Govt. Boys High school hostel	143	10	25	—	178
Govt. Boys total	287	46	216	23	572
GB Middle (6–8)	49	13	25	2	89
GB High (9–10)	134	15	114	12	275
GB Higher (11–12)	104	18	77	9	208
Govt. Girls High (9–10)	134	19	86	20	259
Govt. Girls Higher (11–12)	42	7	31	19	99

Note: Loyola starts from the sixth, SSM goes up only to the tenth, the Government girls and boys school have classes from 6 to 12. The figures for Govt. girls middle school were not available, and nor was the detailed breakup for Loyola English medium school. NA = Not Available.

Table 8.2   Loyola Hostel (2001–02)

Tribals	Non-tribals	Christian	Non-Christian
422	51	409	64

Table 8.3   Saraswati Shishu Mandir (2001–02)

	Christians	Muslims	Hindus	Total
Primary school (1–5)	44 (all tribal)	37	314	395
Middle and High	20	30	356	406
Total	64	67	670	801

The clientele of the RSS school is predominantly lower-middle-class Hindu families, particularly government employees, small business owners, and traders, who can afford the fees of Rs. 80–250 (approximately $2–5) per month (depending on the grade and the locality). In terms of caste, RSS schools have more "other backward castes" (OBC) than adivasis among both students and teachers. The RSS, BJP, and other affiliates have a strong base among the OBCs since they feel threatened by the educational advancement of the Uraons, and are resentful of the reservations they get in government jobs. The Kunkuri Shishu Mandir is somewhat different from other Vidya Bharati schools where the children are almost totally Hindu (Sarkar 1996, p. 245) in that it has approximately 16.4 percent Christian and Muslim students. Most of these Christian adivasi children belong to families with one parent in a government job, since only they can afford the fees. The Kunkuri Viyda Bharati principal claimed that their parents sent them because they were taught discipline and culture in addition to their regular school courses, but this clearly needs further investigation.[18]

In the RSS school, out of a total of 26 teachers, only seven are women, of which one is a Christian Uraon, while the bulk are from the "other backward castes." The minimum qualification for a primary school teacher in the Vidya Bharati network is high school certification, but a little more than half (14 out of 26) are postgraduates. Recruitment is through written exams, followed by an interview, a seven-day training period, and then another exam. Those who complete a year long training course (six months correspondence, six months regular attendance) run by Vidya Bharati at the Saraswati Shiksha Mahavidyalaya in Jabalpur get extra increments. Becoming a teacher in one of the Vidya Bharati schools (unlike many of the students who come to these schools simply to get a good education), generally requires prior and special commitment to the Sangh agenda, which is further reinforced by the ideological training they periodically receive. However, teaching in these schools is also seen as a job—and given that teachers are paid less than the government scales, many of them would be happy to switch to government jobs.[19]

The Government schools are divided roughly evenly between adivasis and "other backward castes" in terms of students. However, approximately only one third of the teachers are adivasis (12 out of 31). Since government schools do not charge fees, unlike the Catholic schools or the RSS school, they tend to attract the poorest adivasi students or those students from business families, particularly girls, for whom schooling is seen as just a way of marking time before they get married or take over the family business. The level of discipline is generally much lower in government schools, with a corresponding decline in pass percentages. Extra curricular activities are also noticeably limited compared with the religious schools. The government boys' school has recently acquired a strict principal, who has ensured better results, leading to a corresponding increase in enrollment.

While pass percentages are an inadequate reflection of school performance, because the results are not weighted by factors such as the numbers enrolled, drop out rates, and the caste, class, and gender composition of students, table 8.4 provides a rough comparison. Both the government schools and Loyola had on average between 100 to 200 students, while the SSM had on average less than 50 students per year.

In sum, while religion evidently does play a role, the ultimate criterion for parents in selecting schools appears to be exam results and cost of schooling. In the public perception, RSS schools fare well on results and are affordable for the lower middle class and thus manage to attract a range of children whose parents are not necessarily committed to the Sangh agenda. Children who graduate from these schools, however, seem to end up with a strong sympathy for the Sangh, especially the discipline it imparts.[20] The following section, based on ethnographic observations in the Kunkuri Shishu Mandir attempts to explain why this is so.

Table 8.4 Tenth grade results: percentage of those who passed the exams

Year	SSM	Loyola	Government girls	Government boys
1991–92	NA	48	22	62
1992–93	NA	63	45	32.4
1993–94	NA	47	15.05	19.6
1994–95	NA	35	6.8	24.3
1995–96	NA	58	24.4	20.4
1996–97	59	66	23.4	51
1997–98	52	49	20.2	25.5
1998–99	51	NA	26.4	59.4
1999–2000	93.84	NA	28.4	52.3
2000–01	NA	67	18.1	47

Note: I have selected the Tenth Board Exam results because this is the only grade for which I have figures for all the schools. The Saraswati Shishu Mandir only started its Xth grade from 1996 to 1997. The percentages have been calculated on the basis of those who sat the exams. NA = Not Available.

SARASWATI SHISHU MANDIR: AN ETHNOGRAPHIC EXPLORATION

The first Saraswati Shishu Mandir was set up in 1952 in Gorakhpur, Uttar Pradesh, although a Gita school had been established by the RSS chief Golwalkar in 1946 at Kurukshetra, Haryana. As the number of schools grew in different states, an all India coordinating body, called Vidya Bharati, was set up with its headquarters in Delhi. The Vidya Bharti educational mission is founded on the objective of training children to see themselves as protectors of a Hindu nation:

> The child is the centre of all our aspirations. *He is the protector of our country, Dharma (religion) and culture.* The development of our culture and civilisation is impact in the development of child's personality. A child today holds the key for tomorrow. *To relate the child with his land and his ancestors is the direct, clear and unambiguous mandate for education.* We have achieve the all round development of the child through education and sanskar i.e. inculcation of time honoured values and traditions. (*sic*) (Vidya Bharati website, 2001; emphasis added)

The Vidya Bharti schools are funded through fees and private donations from rich trading families or other wealthy sympathizers, a point that they make much of when comparing themselves with Church organizations that get money from abroad. In fact, however, as a recent study showed, foreign exchange from nonresident Indian sympathizers of the RSS amounts to significant sums.[21] The RSS also claims to take no government aid, though periods of RSS expansion certainly seem to show an uncanny correlation with having a BJP government in power.[22]

In what follows, I describe different aspects of the school's organizational culture, focusing particularly on the extracurricular markers and makers of identity. To a great and unacknowledged degree, the educational culture of missionary schools has become part of the general educational culture and to that extent, RSS practices may be seen as a perfectly legitimate reaction to "Christian" markers introduced in the guise of "modern" and "secular" education. For instance, in private as well as in government schools, children start with morning assemblies where they are made to pray or subjected to some "uplifting thoughts." While many government schools have started making girls wear the north Indian salwar kameez, boys invariably wear shorts or pants, and in the more elite schools, girls generally wear skirts. Teachers are addressed as "Sir" or "Ma'am" or "Miss."

In the Vidya Bharati schools, however, care is taken to use purely "Hindu" cultural markers. Although the students' uniforms are not dramatically different from those of children at other schools (blue kurtas for girls with white sashes and khaki shorts/pants for boys), in sharp contrast to other schools, teachers here also wear uniforms. Even more unusually, while women have always borne the burden of upholding tradition through sartorial conservatism (in this case white sarees with red borders), here male teachers too flaunt their indigeneity by wearing white dhotis and kurtas unlike the Western pant/shirt that has become universal costume for urban educated men in India. The use of Sanskrit terms to address teachers (Acharya), the practice of touching their feet as a mark of respect, and the naming of classrooms after Hindu sages (Vashisht Kaksh, Vishwamitra kaksh), also mark the school out as a space where Hindu Dharma and Hindu *sanskars* are asserted with pride, where tradition is saved and transmitted as against the "deculturation" or "christanity and Westren [*sic*] mores" that convent schools lead to (Vidya Bharati website). While "sanksars" also refer to the rituals that mark stages of life, they are defined here loosely as good influences, good habits and values, all of which are essential to character formation (Jaffrelot 1996, p. 48).

In addition to the two national days (January 26 and August 15), the Vidya Bharati schools have their own roster of special days to be celebrated, such as the birthdays of Shivaji and Jijabai, Vivekanand, Deen Dayal Upadhyay, and Savarkar, all of whom are icons of Hindu nationalism, or in some cases simply Hindu Mahasabha/RSS leaders. Significantly, Gandhi's birthday, which is a national holiday for the rest of the country, is not celebrated, although in recent years efforts have been made to appropriate him too. Other official holidays are also eschewed, in favor of traditional Hindu referents. For instance, Shikshak Diwas or Teacher's Day (which is officially celebrated on September 5 on the anniversary of the country's former President and well known educationist Dr. Radhakrishnan) is celebrated on the birth anniversary of the Sage Vyasa, which is the traditional Hindu Guru Poornima (Teacher's Day). Krishna Janmashtami stands in for Children's Day, officially celebrated in India on Nehru's birthday, November 14. In the process, myth and history, the birth and death anniversaries of actual historical figures and those of mythical characters are seamlessly conflated and inscribed in the child's consciousness through the regime of annual holidays, celebrations, morning prayers, as well as through the content of history and "cultural knowledge" textbooks.

A similar process of conflation (of real and mythical events/people), elision (of non-Hindu persons and things) and abuse (of facts) is evident in a variety of spheres. The *Ekatmata Stotr*, which students recite to start the school day, lists the names of all the mountains, rivers, Hindu pilgrimage sites, Hindu mythological figures, sacred books such as the Vedas, Upanishads, Jain, Buddhist, and Sikh books, saints and poets, Hindu kings and queens and finally RSS leaders like Hegdewar and Golwalkar. Another example is the Sanskriti Gyan Pariksha primer published by Vidya Bharati at Kurukshetra, from which the question on the Ramjanmabhoomi is reproduced at the start of this chapter. The exam is, on the face of it, a disinterested test of knowledge about the country's geography, history, and culture. It is in question–answer format, and combines sections on actual and mythical Hindu figures, actual events and events from the Mahabharat, Ramayana and so on. RSS fantasies are introduced here as acceptable historical facts, including the notion that Homer's *Illiad* was an adaptation of the Ramayana and that Christ roamed the Himalyas and drew his ideas from Hinduism (Sahmat 2001, pp. 14–18). Needless to say, neither in the Ekatmata Stotr or the Gyan Pariksha Primer, are there references to anything Christian or Muslim and the version of Indian culture that is produced is thus an exclusively Hindu—and more specifically upper caste (mostly Northern)—culture.[23]

More than just a site for the transmission of Hindu cultural traditions, however, the school is portrayed as a temple. Starting with the name itself, Saraswati Shishu Mandir (student temples of Saraswati, the Hindu Goddess of learning), the practice of leaving shoes outside the classroom (as in temples), the chanting of Sangh shlokas in Sanskrit to mark each transitional point such as the break for recess or the end of the school day, the rich visual display of calendar art posters of Hindu Gods and Goddesses[24] are all practices that reinforce the notion of learning as an act of faith or religious devotion that precludes any kind of critical questioning. As against being a space where children can transcend their religious identities and begin to learn about other cultures or develop faculties for critical enquiry, the school becomes an extension of the kind of religious discourse that is imparted in temples. This is not the same as the rote learning for which Indian schools are notorious, but a sense that the functioning of the school helps to keep a religious identity alive.

Textbooks maintain the emphasis on devotional zeal. The history textbook for class V, for example, refers constantly to "Mother India" from whose womb many brave sons were born, who worshipped her and died for her, with the Gita in their hands and *Vande Matram* on their lips (Singh 1997, pp. 27–28). Most of the Vidya Bharati schools are affiliated to the CBSE (Central Board for Secondary Education) or their local State Boards. In general, these schools follow the syllabi (and the textbooks) published by the NCERT. Earlier, there were obvious contradictions between official texts which said that the Aryans came from central Asia and the RSS assertion that they are indigenous to India, or those that give Gandhi pride of place in the freedom struggle and the RSS denigration of Gandhi as a *Dushtatma* (bad soul) for "appeasing" Muslims. Individual teachers may have perfected sophisticated pedagogical techniques for getting around this (Mody 2002), but perhaps what enabled the two belief systems to coexist is the emphasis on teaching in order to get marks and pass exams. The revised NCERT textbooks do much to alleviate this problem, in that they broadly reflect RSS ideology, yet they are still inadequate in that they cannot openly talk about the Sangh.

Vidya Bharati therefore brings out its own textbooks, which "supplement" and "correct" the history that is taught in the official books, working more by selective emphasis on certain figures than by crude propaganda against Muslims and Christians. *Itihas ga Raha hain* (History is Singing) for class V blames "internal disunity" for the invasions by the Turks, Mongols and Mughals, but notes that even in the medieval period the "freedom struggle" was kept alive (Singh 1997, p. 9). While professional historians point to the presence of Hindu generals in Mughal armies and the fact that Shivaji, the archetypal Hindu king had a Muslim general, as evidence of the fact that medieval power struggles can not be understood in religious terms, the RSS sees this as a betrayal of Hindus and reserves its greatest criticism for such "collaborators" (Singh 1997, p. 78). The connection between these views and the violence enacted by the RSS against Hindus who helped Muslims during the 2002 pogrom in Gujarat is clear (Sundar 2002b, p. 117).

Christian pastors are described as one of the main instruments of colonialism (Singh 1997, p. 27), thus strengthening the association in children's minds between Indian Christians and antinational activities. While Gandhi and Nehru are perforce mentioned as leaders of the freedom movement, equal pride of place is given to the Hindu nationalist stream consisting of Tilak, Malviya, and religious men (*sadhus* and *sanyasis*). One of the central planks of the RSS is the equation of "holyland" with "motherland," which supports the claim that because Muslims and Christians have their Meccas elsewhere, they are not fully loyal to the country. The text exhorts children to remember who they are so as not to become slaves again and asks rhetorically:

> Whose is this country? Whose motherland, fatherland and holyland is it? Whose customs and festivals are celebrated according to the agricultural rhythms and climate of this land?[25] Which people is it who call Sivaji, Ranapratap, Chandragupta, Bhagwan Ram, Krishna, Dayanand their great leaders.

It then goes on to a fervent description of the greatness of the RSS founders, Hegdewar and Golwalkar, and the need for an organization such as the RSS to build Hindu unity (Singh 1997, pp. 77–81).

In addition to the prescribed curriculum, Vidya Bharati schools teach five extra subjects all of which are thought to contribute to the development of *sanskar* or character formation: moral education (*naitik shiksha/ sadachar*), which includes stories about great men, songs, instruction on honesty, personal hygiene and so on; physical education (*sharirik shiksha*), which includes learning to wield a stick, dumbbells, and martial arts; *Yoga*; music/singing (*sangeet*); and Sanskrit (from kindergarten onward and not just from the third as in government schools). "Vedic Mathematics," alleged to be indicative of the mathematical prowess of the Vedic peoples but which contemporary Indian scientists describe as nothing more than a computational tool with no Vedic origins, is introduced in the third standard.

While the schools attempt to inculcate Hindu *sanskars*, they also borrow liberally from elite school organizational forms. Students are organized into groups (Bal Bharati for classes 6–8; Kishore Bharati for classes 9–11 and Kanya Bharati for girls) to teach them self governance. These groups are assigned responsibility for the library, singing practice, celebration of national days, and so on. During the hour and a half long Kanya Bharati sessions per week, girls learn sowing and cooking so that, as one teacher explained, they can be good housewives and mothers "like Jijabai (who exhorted her son Shivaji to war) and Lakshmibai (who strapped her child on her back and fought the British in 1857)."[26]

In order to achieve more complete indoctrination, a constant attempt is made to wrest control over the family's socialization functions as well. Unlike the usual practice of calling parents to school for parent–teacher interactions, teachers from the Saraswati Shishu Mandirs go to students' houses and tell their parents about the children's performance, activities planned by the school, and other matters. They also have an *abhibhavak sammelan* or gathering of parents once a year. Children also serve as messengers by taking home pamphlets about RSS or Vishwa Hindu Parishad activities for their parents. Teachers keep an eye out for promising students or volunteers who can go on to RSS summer camps known as Officer Training Camps and receive further ideological training. According to the principal of one of the Shishu mandirs, about 2 percent of the students from middle and high school go to these camps every year.[27]

If Vidya Bharati schools stopped at transmitting a religious identity per se, they would be little different from Catholic schools, whose daily routine is also marked by morning and evening mass, which is compulsory for Catholics but not for non-Catholics. What marks the transition to fascist schooling, however, is the overriding emphasis on the physical "defence" of religion. At the end of the school day, following the *Visarjan mantra*, the teacher asks *Hamari Mata kaun hai* (who is our mother?), to which the students reply, hands raised, fists clenched: *Bharat Mata, Bharat Mata* (Mother India). The teacher then asks *Iski raksha kaun karega, kaun karega* (who will protect her, who will protect her?), to which they shout *Hum Karenge, Hum Karenge* (we will, we will).

At Republic Day (January 26) celebrations in a Saraswati Shishu Mandir school in Raipur in 2002, the same slogan was repeated after the national flag was hoisted. The rest of the program was a long succession of speeches on terrorism, self-defense, and the need to fight Pakistan, thus inscribing blood and guts as the very essence of patriotism. For instance, the female student (leader of Kanya Bharati) who was conducting the program said that the 24 spokes in the Ashok chakra in the center of the Indian flag "reminds us that we should be 24 hours alert in the defence of the country." After a display of physical exercises, *lezium*, gymnastics, and yoga, honoring the country's martyrs and Hindu *dharm*'s "four gods and four castes," primary school children sang a song about the need to fight the "neighboring country" and demolish it as brave children of Savarkar (Hindu Mahasabha leader), a theme that was reiterated by the Chief Guest's address at the end of the program. In between was a speech in English on "terrorism," another one by a fifth grade girl in Hindi on September 11 and December 13, which pitted the "conservative, fundamentalist, terrorist Muslim community" on one side versus the "powerful American community" on the other. India lay in between, with its belief in *vasudeva kutumbkam* (the whole world is one family). A high school boy gave a speech in Chhattisgarhi on "Defence" and the "danger for Hindus" faced with enemies both outside and inside the country (i.e., Muslims). One very small girl talked about how the constitution laid down rights and duties for all, but some people only thought of their own rights and did "injustice" to the rights of others.

On the face of it this was an unexceptional exercise in reminding people of their civic duties, but in the light of the RSS discourse on government "appeasement" of minorities at the expense of the "majority," acquires a clear anti-Muslim and anti-Christian bias. Indeed, the fact that Pakistan and Indian Muslims are never explicitly named but referred to only as "the neighboring country" or "terrorists" paradoxically serves to strengthen the message, since the RSS can then take recourse to the claim of pure "nationalism." Indeed, students claim they are taught only about "Indian" culture and deny that they are taught to hate other communities.[28]

Interspersed between the speeches were catchy songs describing the victory of Hindu kings like Rana Pratap against the Mughals or comparing Hinduism to a sea into which all religions flow. The Sangh insists that all those living in India are part of "Hindutva," which it defines as a "way of life" and simultaneously demonises Christianity and Islam for being foreign to India. It thus claims both Hindu tolerance and the need to defend Hinduism against the very existence of religions of which it needs to be tolerant. The entire program was prefaced and suffixed with an *aarti* (lighting of lamp, garlanding) to a picture of "Mother India" (Durga against a map of India, representing a Hindu nation).

While Republic Day is inevitably an occasion for a display of patriotism, a moment when schooling reveals its connections to state projects, and the constitutional ideal of secularism and sovereignty is reiterated (however imperfect the practice), in the hands of Vidya Bharati, such occasions become weapons to destroy this fragile constitutional existence. If children are schooled, it is only to become better soldiers for a Hindu nation. Even the leaders of the groups into which students are divided for extracurricular activities are called *senapatis* (generals). For the RSS, as for the Nazis, "education is never for its own sake; its content is never confined to training, culture, knowledge, the furtherance of human advancement through instruction. Instead it has sole reference, often enough with implication of violence, to the fixed idea of national pre-eminence and warlike preparedness" (Mann 1938, p. 6; see also Kandel 1935, pp. 12–13).

Conclusion

This chapter looks at schooling as part of the RSS agenda to create certain notions of citizenship and identity. The RSS/BJP has attempted to affect a radical departure in the existing educational ethos, through the use of both state power (packing state educational institutions with its own ideologues), and the instruments of "civil society" (creating its own network of schools in order to feed into a well developed cadre structure). Like revolutionary movements elsewhere, the RSS believes in the idea of education as a means toward social transformation. However, whereas Soviet communists, at least in theory, saw education as a means toward liberation, equality and inclusion (see Ewing 2004), the RSS's aims are indoctrination, hierarchy, and exclusion.

Inevitably, the "success" of ideological reproduction through schooling depends on a variety of factors such as the degree of conformity of the child's lived experience with the dominant ideology, or the extent to which this ideology is reinforced through wider cultural practices and political developments in society. So far, children going to RSS schools have tended to come from upper-caste Hindu backgrounds and schooling has merely reinforced existing sympathies,[29] but as the RSS expands among dalits and adivasis, their relationship with the Sangh's message is bound to be somewhat different. The RSS educational and political agenda includes both absorbing subaltern groups into a Hindu fold to fight against "minorities" and using violence against these same groups in order to perpetuate the existing social order.[30]

RSS schools are popular with a wider circle of parents beyond the strictly converted, primarily because they perform the requisite *educational* function of producing "good exam results," which appeals to middle-class parents and children in a certificate- and degree-oriented economy in a context where state schools suffer from gross neglect and underfunding. In keeping with Gramsci's idea that the content of education must be seen to be disinterested to become hegemonic (Forgacs 1988, pp. 313–318), while such non-RSS parents may not desire an overtly Hindu education, here the discipline and *sanskars* become a bonus because they are tagged on to success in exams. The positive valuation of discipline is also perhaps a result of the widespread internalization of the Sangh's argument that Hindus were historically vulnerable and defeated by invaders because they lacked unity and organization, compared with Muslims and Christians, and that they need to be disciplined to defend themselves.

The situation also varies across genders. In a society where films and social custom privilege certain modes of womanhood—the idea that girls must study and perhaps even have careers but never abandon their primary duty to their family—girls themselves seem to welcome the training in "values" they get.[31] The Sangh's appeal lies in its ability to lay claim to Hindu culture at large, to conceal its own warped and petty version of Hinduism within a larger stream of diversity.

What non-RSS parents need to realize, however, is that the RSS project is distinct from all other schools in the harnessing of pedagogy to a clear political end. Although in the past Jesuit schools had great proselytizing zeal, now schools like Loyola act primarily as channels for middle-class mobility. RSS schools, while teaching children to pass exams and get white-collar jobs, also produce children, particularly boys, steeped in aggression and chauvinist attitudes toward non-Hindu "others." With boys, this aggression is also directed to girls and women, with the nonconventional or non-Hindu woman becoming an object of both prurient interest and disgust. Curricular and extracurricular messages such as uniforms, functions, or cultural knowledge exams all serve to remove non-Hindus from the discursive space of the nation. When they appear it is only as insoluble "problems" for the Hindu body politic. Coupled with the emphasis on militarism, such as physical training in knife and stick wielding and repeated exhortations to "defend" the "nation," the stage is set for internal civil war. The tragedy is that, imbricated in the banality of exams and results and the middle-class desire for service jobs, this is legitimized as just an alternative form of education.

AFTERWORD

Since this article was written (December 2003), the national situation has changed considerably. The BJP led coalition, the National Democratic Alliance, lost the general elections in May 2004, and was replaced by a Congress led coalition government, the United Progressive Alliance (UPA). The new government set up a committee to review the NCERT textbooks and has appointed a new Director for the NCERT, who is a well-known and progressive educationist. The national situation thus appears more hopeful. However, the RSS continues to run its own schools, and influence policy in the states where the BJP is in power, such as Chhattisgarh. While the first half of the article is dated, it shows what might happen were the BJP to return to power. The second half, which describes the Saraswati Shishu Mandirs, continues to be of salience.

NOTES

I am grateful to Krishna Kumar for suggesting ways in which this research could be done. I thank Tom Ewing, Tanika Sarkar, the participants at the workshop on Revolution and Pedagogy, 2001, Ohio State University and audiences at the Cambridge, Edinburgh, and Oxford South Asia Centres for their comments.

1. The Rashtriya Swayamsevak Sangh (RSS), an ostensibly "cultural" organisation was set up in 1925 by Dr. K. B. Hegdewar to promote a "Hindu rashtra" or Hindu nation in which members of minority religions would occupy a subordinate role. The RSS has no official membership but has an estimated 45,000 *shakhas* or branches all over the country (Noorani 2000, p. 13), and a well developed organizational structure. *Shakhas* are places where boys and men, wearing the standard RSS uniform of khaki shorts, meet for an hour a day to do physical exercise and military style drill which includes training in martial arts, and learning to wield a stick or a knife, ostensibly for "self defence." They also receive instruction in proper behavior (*sanskars*) and ideological training (*baudhik*) (Anderson and Damle 1987; Hansen 1999, p. 113; Jaffrelot 1996, pp. 35–40; Kanungo 2002, pp. 25–26). The RSS was banned in 1948 after one of its (allegedly former) members killed Mahatma Gandhi for ostensibly being pro-Muslim, and again during the Emergency (1975–77), when normal rights were suspended and people were arrested across the political spectrum. For long, it was a secretive and shadowy organization—for instance, its finances are still not audited. Since the BJP came to power, it has achieved public respectability and gets considerable and positive media coverage.

2. For a history of the scholarly debate over the site, see Lal (2003, pp. 141–185). In September 2003, the government run Archaeological Survey of India issued a report concluding that the mosque did indeed replace a temple. University archaeologists have described this report as biased and unprofessional, pointing out several flaws in method and interpretation (Sahmat 2003).

3. This article was completed in December 2003, when the BJP was still in power. See the afterword for an update.

4. The University Grants Commission funds and administers all universities in the country. Other national educational bodies which have been packed by RSS sympathizers (almost all unknown figures in their professional fields) include the

Indian Council for Social Science Research (the Indian equivalent of the Social Science Research Council), and the Indian Council for Historical Research.

5. <www.Vidya Bharati Akhil Bharatiya Shisksha Sansthan.htm>.

6. Since the term *adivasi* challenges the notion that the Rig Vedic peoples were the original settlers of the country, the RSS prefers to call them *vanvasis* or forest dwellers.

7. In central India, Sewa Bharati runs a girls' hostel in Pathalgaon and two boys' hostels in Gwalior and Bhopal. They also have a big school at Kurukshetra near Delhi. According to one VKA spokesperson, the idea behind Sewa Bharti's schools was to train future administrators.

8. The institutional contexts are clearly different but the conservatives' attempt to tar professional historians as antinational and unpatriotic, when they challenge dominant histories, is remarkably similar. For instance, in 2001 when professional historians in India protested against deletions in existing texts that ostensibly hurt religious sentiment and objected to the idea that religious leaders should vet textbooks, the Human Resource Development Minister in India responded by calling it "intellectual terrorism unleashed by the left . . . more dangerous than cross border terrorism" (*Indian Express*, December 20, 2001). Such language would not be unfamiliar to those involved in the American debate over the National History Standards. A letter from Kim Weissman in the *Wall Street Journal*, for example, accused the authors of the National History Standards of wanting to "indoctrinate children with their own hatred of America; to steal the American birthright from the children of our country; to teach our children to feel guilt over their own heritage. Now the special interest pressure groups seek, through Goals 2000, to complete the balkanization of America" (reproduced in Nash et al. 2000, p. 189).

9. Militarized education can be found in other context, such as the pre–World War II British public schools. For examples in contemporary Indian education, see Benei (2003).

10. There were brief exceptions such as in the 1970s when four history textbooks were withdrawn by a right leaning government (Chaudhury 1977).

11. For instance, the government boys school in Kunkuri celebrates Saraswati Puja, in honour of the Hindu goddess of learning. Much depends on the predilections of individual principals and teachers.

12. This attempt to "Indianize" at the university level includes introducing courses like Vedic rituals and Vedic Astrology. Many of the country's scientists and social scientists repudiate the latter as spurious science, and not particularly Indian. An appeal against the course is currently pending in the Supreme Court (Sundar 2002a).

13. Skocpol defines social revolutions as "rapid, basic transformations of a society's state and class structures . . . accompanied and in part carried through by class-based revolts from below . . . set apart from other sorts of conflicts and transformative processes above all by the combination of two coincidences: the coincidence of societal structural change with class upheaval; and the coincidence of political with social transformation" (Skocpol 1979, p. 4).

14. Judgement by Justice M. B. Shah, D. M. Dharmadhikari, and H. K. Sema in Writ Petition (Civil) No. 98 of 2002, Ms. Aruna Roy and others versus Union of India and others.

15. See Habib et al. (2003), and the riposte by Lal et al. (2003), which brings the debate to a new low. See also Sahmat 2002.

16. For similar conclusions about the role of schooling in generating feelings of inferiority and thus reproducing structures of class, gender, and ethnicity elsewhere, see McCarthy and Crichlow (1993) and Brown et al. (1997).

17. Currently, the paucity of buildings forces it to run in two shifts—7.30 AM to 11.00 AM for kindergarten and grades 1–4 and 11.30 AM to 4.30 PM for grades 5–10.

18. A more likely reason is that Loyola starts only from middle school and the Loyola English medium is slightly more expensive.

19. They get between Rs. 1250 and Rs. 1800 ($26–38) per month as basic pay, depending on whether they are primary or higher secondary school teachers. The managing committee for each locality decides on how much it is able to pay.

20. Interviews with graduates of the Saraswati Shishu Mandir, studying in the government girls' high school and in Loyola school.

21. South Asia Citizens Web and Sabrang Communications (2002).

22. Although the RSS organizations claim they refuse to take government aid on principle because it saps their voluntarism and makes them dependent on the whims of the government of the day, access to state power through the BJP has definitely helped in expanding the scope of schools and other organizations. The Principal of the Saraswati Shishu Mandir in Rohinipuram, Raipur said that although Vidya Bharati work began in Chhattisgarh in 1968, they received a new impetus in the 1990s (coinciding with the BJP-led Sundarlal Patwa government). Reports of government patronage—gifts of prime land in the Capital to RSS organizations at very low rates—made newspaper headlines in late 2002.

23. This both reflects and reinforces a phenomenon in wider cultural practice and media representations as well, where a Muslim or Christian presence in films or elsewhere is much more rare than it used to be.

24. Unlike government schools whose walls are bare, save perhaps one poster of Saraswati or Ganesh over the blackboard, or Catholic schools which have a cross in each classroom, the Saraswati shishu mandirs have a proliferation of calendars and posters with pictures of Gods/Goddesses. Ram in his martial pose is particularly popular.

25. The implication clearly is that festivals like Id and Christmas are not locally rooted.

26. Interview with teachers at a Saraswati Shishu Mandir in Raipur.

27. For the importance of camps to fascist education, see Schiedeck and Stahlman (1997)

28. Interview with ex-students of SSM, now at the Government Girls higher Secondary School, Kunkuri

29. Tanika Sarkar (1996, p. 246) notes that "with school hours in daytime and shakhas in early morning or evening, the whole day is disciplined by various agencies of the sangh, stretching from family to school and shakha, in an unbroken and totalitarian circuit of influence and training. The schools, with a committed lot of teachers and a good record of discipline also act as pivotal points of RSS influence within entire localities, where they function in conjunction with other RSS facilities."

30. In Gujarat, dalits and tribals were mobilised by the Vishwa Hindu Parishad (VHP) to loot, burn, and kill Muslims, during the genocide of March 2002. In October the same year, however, in Jhajjar, Haryana, a VHP mob lynched five dalits for allegedly skinning a live cow, followed by one of their leaders declaring that the life of one cow was more important than that of five dalits.

31. Interview with ex-students of SSM, now at the Government Girls Higher Secondary School, Kunkuri.

REFERENCES

Anderson, W. and S. Damle. 1987. *The Brotherhood in Saffron: The Rashtriya Swayamsevak Sangh and Hindu Revivalism.* Delhi: Vistaar Publications.

Benei, V. 2003. "Military Schools as Hybrid Political Projects: Nehru, Gandhi and the Hindu Nationalists." Unpublished paper.

Blackburn, G. W. 1985. *Education in the Third Reich: A Study of Race and History in Nazi Textbooks.* Albany: SUNY Press.

Brown, P., A. H. Halsey, H. Lauder, and A. S. Wells. 1997. "The Transformation of Education and Society: An Introduction." In *Education, Culture, Economy, Society.* A.H. Halsey, ed., Oxford: Oxford University Press.

Chaudhury, V. C. P. 1977. *Secularism vs. Communalism: An Anatomy of the National Debate on Five Controversial History Textbooks.* Patna: Navdhara Samiti.

Delhi Historians' Group. 2001. *Communalisation of Education: The History Textbooks Controversy.* Delhi: Delhi Historians' Group.

Ewing, E. Thomas. 2004. "Gender Equity as Revolutionary Strategy." This volume.

Forgacs, D. ed. 1988. *An Antonio Gramsci Reader: Selected Writings 1916–1935.* New York: Schocken Books.

Furer-Haimendorf, C. Von. 1982. *Tribes of India: The Struggle for Survival.* Delhi: Oxford University Press.

Giles, Geoffrey J. 1992. "Schooling for Little Soldiers: German Education in the Second World War." In *Education and the Second World War: Studies in Schooling and Social Change.* Roy Lowe, ed., London: The Falmer Press, pp. 17–29.

Golwalkar, M. S. 1939. *We, or Our Nationhood Defined.* Nagpur: Bharat Prakashan.

Habib, Irfan, Suvira Jaiswal, and Aditya Mukherjee. 2003. *History in the New NCERT Textbooks—A Report and an Index of Errors.* Kolkata: Indian History Congress.

Hansen, T. B. 1999. *The Saffron Wave: Democracy and Hindu Nationalism in India.* New Delhi: Oxford University Press.

Hein, Laura and Mark Selden. 1998. "Learning Citizenship from the Past: Textbook Nationalism, Global Context and Social Change." *Bulletin of Concerned Asian Scholars* vol. 30, no. 2, pp. 3–17.

Hellinger, Daniel and Dennis R. Judd Brooks. 1991. *The Democratic Facade.* Cole Publishing Company.

Indian Express. December 20, 2001.

Jain, Meenakshi. 2002. *Medieval India: A Textbook for Class XI.* New Delhi: NCERT.

Jaffrelot, C. 1996. *The Hindu Nationalist Movement in India.* New Delhi: Penguin Books.

Jeffery, Roger, Patricia Jeffery, and Craig Jeffrey. 2002. *Privatisation of Secondary Schooling in Bijnor: A Crumbling Welfare State?* (unpublished paper).

Kandel, I. L. 1935. (rpt. 1970). *The Making of Nazis.* Westport: Greenwood Press.

Kanungo, P. 2002. *RSS's Tryst with Politics.* Delhi: Manohar Publishers.

Kumar, K. 2001. *Prejudice and Pride: School Histories of the Freedom Struggle in India and Pakistan.* New Delhi: Viking.

Lal, Makhan, Meenakshi Jain, and Hari Om. 2003. *History in the New NCERT Textbooks: Fallacies in the IHC Report.* New Delhi: NCERT.

Lal, Makhan, Meenakshi Jain, and Hari Om. 2002. *India and the World: Social Sciences Textbook for Class VI.* New Delhi: NCERT.

Lal, Vinay. 2003. *The History of History.* New Delhi: Oxford University Press.

Luykx, A. 1999. *The Citizen Factory: Schooling and Cultural Production in Bolivia.* Albany: SUNY Press.

Mann, T. 1938. *School for Barbarians*. New York: Modern Age Books.

McCarthy, C. and W. Crichlow, eds. 1993. *Race, Identity and Representation in Education*. New York: Routledge.

Menon, Usha. 2003. "Where Have the Mangoes Gone?" *Economic and Political Weekly*, May 3, pp. 1747–1749.

Minio-Paluello, L. 1946. *Education in Fascist Italy*. London: Oxford University Press.

Mody, A. 2002. "Manufacturing Believers." *The Hindu*, February 10.

Nambissan, G. 2000. "Identity, Exclusion and the Education of Tribal Communities." In *The Gender Gap in Basic Education*. R. Wazir, ed. New Delhi: Sage Publications, pp. 175–224.

Nanda, B. 1994. *Contours of Continuity and Change: The Story of the Bonda Highlanders*. New Delhi: Sage.

Nash, G. B., C. Crabtree, and R. Dunn. 2000. *History on Trial: Culture Wars and the Teaching of the Past*. New York: Vintage Books.

NCERT 2000. *National Curriculum Framework for School Education*. Delhi: NCERT (National Council for Educational Research and Training).

Nelson, John K. 2002. "Tempest in a Textbook: A Report on the New Middle-School History Textbook in Japan." *Critical Asian Studies* vol. 34, no. 1, pp. 129–148.

Noorani, A. G. 2000. *The RSS and the B.J.P.: A Division of Labour*. New Delhi: Leftword Books.

PROBE. 1999. *Public Report on Basic Education in India*. New Delhi: Oxford University Press.

Sahmat. 2001. "Communalisation of Education and Culture: Sahmat Statement." In *Against Communalisation of Education: Essays, Press, Commentary, Reportage*. New Delhi: Sahmat and Sabrang.com.

———. 2002. *Communalisation of Education: The Assault on History: Press Reportage, Editorials and Articles*. New Delhi: Sahmat.

———. 2003. *Against Communalisation of Archaeology: A Critique of the ASI Report*. New Delhi: Sahmat.

Sarkar, T. 1996. "Educating the Children of the Hindu Rashtra: Notes on RSS Schools." In *Religion, Religiosity and Communalism*. P. Bidwai, H. Mukhia, and A. Vanaik, eds. Delhi: Manohar Publishers.

Schiedeck, Jurgen and Martin Stahlman. 1997. "Totalizing of Experience: Educational Camps." In *Education and Fascism: Political Identity and Social Education in Nazi Germany*. Heinz Sunker and Hans-Uwe Otto eds. London: The Falmer Press. pp. 54–77.

Singh, Rana Pratap. 1997. *Itihas Ga Raha Hain, Part II, Textbook for Class V*. Patna: Shishu Mandir Prakashan.

Skocpol, T. 1979. *States and Social Revolutions*. Cambridge: Cambridge University Press.

Sundar, N. 2002a. "Indigenise, Nationalise and Spiritualise: An Agenda for Education?" *International Social Science Journal* vol. 173, September, pp. 373–383.

———. 2002b. "A License to Kill: Patterns of Violence in Gujarat." In *Gujarat: The Making of a Tragedy*. Siddharth Varadarajan, ed. New Delhi: Penguin, pp. 75–134.

South Asia Citizens Web (SACW) and Sabrang Communications. 2002. *A Foreign Exchange of Hate*. New Delhi: Sabrang Communications.

Vasavi, A. R. 2000. *Exclusion, Elimination and Opportunity: Primary Schools and Schooling in Selected Regions of India: Summary of Field Research* (Unpublished paper).

Vidya Bharati website. 2001. <www.Vidya Bharati Akhil Bharatiya Shisksha Sansthan.htm>.

Wolff, Richard J. 1992. "Italian Education During World War II: Remnants of Failed Fascist Education, Seeds of the New Schools." In *Education and the Second World War*. Roy Lowe, ed. London: The Falmer Press, pp. 73–83.

Three Revolutions and an Afterword

Margaret A. Mills

The seminar from which this collection of essays derived was born in the observation of three ideologically distinct revolutions that engulfed adjacent Persian-speaking countries—Iran, Afghanistan, and Tajikistan—between 1978 and 1994. In each case, the political upheaval was accompanied and/or driven by transformations in cultural aspirations and identity negotiations, and major changes in information institutions, especially formal and informal education. In Tajikistan in the early 1990s, a civil war involving regional and religious factionalism, little publicized in the West, followed the dissolution of the Soviet Union, killed perhaps 40,000 people, and created a wave of refugees into adjacent countries, including still war-worn Afghanistan. Its eventual resolution left an occupying contingent of the Russian army in the country for several years. The country is struggling to keep its educational infrastructure economically viable, with government salaries for some university faculty in the humanities still hovering around $10–15 per month, others slowly rising to a still nominal $40–50/month, in the aftermath of the collapse of state revenues that ensued with the end of the Soviet system. Nongovernmental organizations (NGOs) such as the Aga Khan Humanities Project for Central Asia, at the postsecondary level, and the Soros Foundation/Step by Step, at the elementary level, now seek to develop new pedagogies and new curriculum materials to meet the needs of civic education for popular engagement in a fragile post-Soviet democracy.

In Iran, the 1978 revolution engaging a broad spectrum of political factions from secularizing liberal to Marxist to Shi'a religious revivalist, devolved within two years into Iran's first theocratic government. This ideologically revolutionary regime replaced a minimally ideological, highly repressive, authoritarian secular-nationalist, petro-economics-driven monarchy. Early effects on Iran's already well-developed public education curriculum included strengthening existing religious studies and Arabic teaching through the secondary level, curtailment of certain subjects (e.g., Western-style social sciences) at the postsecondary level, and a curtailment of women's access to certain professions served by postsecondary studies, as varied as judgeships and field geology. Twenty-five years later, Iranian general education has produced a solidly majority-literate population with a large urban contingent of citizens under age 30 who are global media-savvy. These trends in information access support increasing popular pressure for a more representative elected government and a reduction of the power of the national religious council, that presently holds the veto over legislative initiatives and the eligibility of candidates for election.

In Afghanistan, a Marxist coup in the capitol in 1978 launched within the next two years aggressive attempts to extend basic literacy and political indoctrination throughout a 75+ percent illiterate (and 85 percent rural) population. Schools immediately became a main arena of contestation between revolutionary and counterrevolutionary forces. Women (of whom very few were employed outside the home) were a particular target of the Marxist government's adult literacy campaigns. Those programs in turn were a particular flashpoint of counterrevolutionary resistance, as they were portrayed as compulsory (despite the government's expansively advertised but actually limited capacity to deliver them), aggressively antireligious in curriculum content, and invasive upon the religiously mandated household authority and protective responsibilities of men. As the armed resistance consolidated during the 1980s, ramped up by large infusions of American, Saudi, and other anti-Soviet third-party armaments and cash, government schools in the countryside were destroyed and government teachers warned off or killed in *mujahiddin* (religious resistance fighter) attacks, not because of any popular rejection of formal education as such, but specifically rejecting the ideological program of the government.

Two dozen or so women teachers I interviewed in 1994 and 1995 in urban Herat, in western Afghanistan, following the departure of the Soviets in 1989 and the defeat of the surviving Marxist regime, had received their teacher training in the Marxist 1980s. Some who were already teaching during the Marxist period described how they had conspicuously carried plastic buckets with shampoo and other bath paraphernalia, pretending to be going to the women's public baths, each morning when they left their homes for their teaching jobs in nearby schools, and how they entered and left their schools as surreptitiously as possible, in order to avoid *mujahiddin* harassment when that presence was most felt in the city. At the same time, anecdotes and life history narratives of the Marxist period include descriptions of an atmosphere of fear engendered by teachers' attempts to enlist schoolchildren to report counterrevolutionary conversations in their homes, as well as accounts of forced selection of children for study abroad in Soviet countries. In 1994, I was told by the director of the government Teachers' College in Herat that the College had run with full enrollments throughout the Marxist period, and that women had comprised the majority of students during the years when the government and the resistance combined put the most severe pressure on the eligible male population, through army drafts (including indiscriminate street impressments) and intensive recruitment of male refugees by the resistance.

By the mid-1990s, a cadre of Marxist-government-trained, but not visibly leftist, women teachers were employed, along with the available men, in schools that were severely underfunded but bursting at the seams, with classes held in hallways and in tents on playgrounds, and elementary grades running three sessions a day in the city schools. Afternoon sessions in some girls' schools were commandeered to accommodate more boys, as the female school administrators complained. Up to one-third of the Afghan population had been external refugees for some substantial period between 1978 and 1994, with an additional one-sixth as internal refugees in government-held cities. Some rural refugees had better access to both schooling and health care in refugee camps and urban residence areas in Pakistan and Iran than they had had in peacetime Afghanistan. In the 1990s, both returned-refugee and stayed-on families were avid to get their children an education, regarded as especially critical for boys, for better employment-market access within and beyond Afghanistan's shattered economy. Their pro-education orientation was economic, but also reflected a general feeling, articulated to me by both male and female adult literacy class participants, that education is necessary for all Muslims, both male and female. It is in fact so mandated by at least two well-known *hadith* (authoritative sayings of the Prophet): "Seek knowledge, even though it be in China," and "Seeking knowledge is the responsibility of all Muslim men and women." The latter was inscribed in Arabic and in Persian translation on posters produced by the Muslim Sisters/Afghanistan organization.

This religious development motive cannot be detached from people's pervasive sense of the urgent need for postwar cultural identity recovery. The goals of the Marxists were perceived to have included the erasure of Afghan national identity as well as political autonomy. This national identity theme is a remarkably vigorous general sentiment despite Afghanistan's historically weak central government and sometimes-intense internal ethnic factionalisms. Afghanistan's current troubles are all but universally, and not illogically, attributed to outside interference and to the misguided collusion of some self-interested Afghans in such interference. Whatever else it is, though, for most Afghans, to be Afghan is also to be Muslim, perhaps observant, perhaps not. The thirst for schooling and other institutions controlled by "Afghans" was not separate from this religious concern.

Throughout the anti-Soviet period of the war, some *mujahiddin* parties maintained ambitious educational programs, and certain international NGOs as well as USAID and other international aid agencies (e.g., the Swedish Committee of the Red Cross) had active programs to support nongovernment schools in Afghanistan as well as in refugee camps. By the mid-1990s, UNICEF, certain NGOs' and European nations' overseas aid agencies were active in rebuilding Afghan schools, rural and urban. Lack of a living wage for teachers in the absence of an effective central government was proving to be a major obstacle to public aspirations for revival and expansion of the education system.

Thus "reactionary" is too simple a term to describe the attitudes toward schooling in the Afghan counterrevolution, much less to trace the two revolutions' impact on education, material or aspirational. The intense desire for reconstruction of the education system after the Marxists' departure was both pragmatically and ideologically driven. The notorious clamp-down on public education by the Taliban in 1995–2001, after their takeover of major

Afghan cities, especially the closure of girls' schools and dismissal of women teachers, was an ideological decision, necessitated, they said, by the lack of suitably Islamic teaching materials in appropriate languages, and of public security for the operation of schools especially for girls. One of its effects, and possibly a tactical intention, was also to impose and demonstrate regime control over would-be resistant men, through threats to the women and girls under their protection. School closure worked an immediate material hardship on women teachers and their families, who depended on their wages for cash income, however slim. The clandestine girls' schools that received international media attention during the Taliban years were, according to anecdotal interview information I was able to gather in Kabul and Herat in 2003, if anything substantially more widespread in the cities than international reporters had suggested. Women teachers that I met included some who were now again students, in reopened university classes. Their comments reflected a mix of ideological and material concerns that drove their decisions to conduct home schools. Some emphasized, others discounted the physical risk involved. Some subscribed to the heroic-resistance narrative, describing the teaching of girls as part of their national and/or religious duty over against the misguided or mischievous, even "anti-Afghan" policies of the Taliban. Others simply said that they taught because their families needed the income from private students' modest fees, and they had to keep working somehow.

MIXED AGENDAS, COMMON PATTERNS

While much more research needs to be done by Afghans and others to construct an adequate (of necessity, largely oral) history of the last 25 years of social upheaval in Afghanistan, the as-yet un-assessed scope and mixed agenda of the girls' clandestine schooling effort alone illustrate a common feature of all the revolutionary efforts traced in this volume. Mixed agendas and differential commitments unfold, between the ideal and the pragmatic, whether dedicated to ideology and identity work or to optimizing pragmatic relations with existing and evolving configurations of power. Regarding the mixed effects of ideologies and opportunisms, Esmail Nashif's Palestinian case study makes explicit some limits of Louis Althusser's distinction between the ideological/hegemonic and the repressive state, in parallel with other discussions here of the ideological and pragmatic that do not engage Althusser so directly. As Nandini Sunder describes the BJP/RSS's concerted effort to construct an ideologically persuasive school and out-of-school instructional system to promulgate a racialist and fascist Hindu nationalism, she also observes parents' pragmatic, nonideological reasons for entrusting their children to such schools, in the quest for skills to empower their children in highly competitive higher education and job markets. She leaves open, but threatening, the prospect of the students' assimilating an ideology that their parents in some measure ignore.

Violent physical rejection of a proffered ideology, so visible in the *mujahiddin* attacks on the Marxist-led school reform in Afghanistan, is easier to trace than is indifference as a form of resistance. The women teachers of Afghanistan who trained in the Marxist era, and taught when and how they could in succeeding years, in and out of schools, expressed no allegiance to the leftist idealism that provided them a greatly expanded access to teacher training, but they certainly show a commitment to the idea of schooling for women, itself a potentiating revolutionary change in Afghanistan's public sphere. Public female schooling has developed in fits and starts, expanding and contracting in secularizing and in religious contexts, in Afghanistan since the early twentieth century. In Afghanistan as in the context of nationalist Egypt in the early postcolonial period, as discussed by Barak Salmoni and in the early Soviet Union in E. Thomas Ewing's chapter in this volume, the argument for physical investment in girls' schooling implies at the practical level redeployment of up to 50 percent of state and nonstate education resources, while these initiatives also "shake [ideological] foundations," in Ewing's words. Those foundations are the entrenched social values, perhaps pre- or sub-ideological as Nashif's case analysis suggests, of patriarchal domination. If gender and sexuality issues are discounted as "red herrings" of conservatism or presented as supportive, even "minor" subtexts to other, more front-stage ideological projects, such back-staging may be tactical (e.g., taking a gradualist, oblique rather than a head-on tactical approach to one of the fundamental and pervasive constraints on human rights equity and human resource development worldwide). Both Salmoni and Ewing describe how increased female access to education, whether segregated or coeducational, was advocated for its anticipated beneficial effects on *men* as members of family units and of society in general. While this may well indicate an instrumental, rather than a primary, value put on expanded female education within the ideology in question, it also reminds us that the unit of social analysis must be kept in view; the individual (male or female) in liberal formations, the family or other collective in other ideologies. The ambiguities of access or equity arguments and critiques, whether or not focused on gender, may be taken as evidence of ideological "woolly thought," or they may be seen as strategic opportunism in the mixed ideological environments of revolutions in progress.

Nandini Sunder's observation of the BJP/RSS's ideological exploitation of young men's potential for aggression toward women and girls, their targeting of *female* representatives of the "other" (religious, cultural) is illustrative of such opportunism in a negative mode, and troubling. The pervasive success of the Congress Party against the BJP in India's 2004 elections may strengthen the hand of policy advocates intent on shoring up India's institutional commitment to secular constitutional democracy, in school curricula as well as other institutions. Any change in the gender politics that impinge on Indian women's educational and public social participation may be harder to trace or predict, in part because the strategic targeting of females in control-politics is seen as a side effect rather than a deep pre- or sub-ideological formation.

Overall, BJP/RSS success in ideological recruiting through their schools over the last decade is not yet ripe for assessment, nor is that effort over.

One telling factor, as with the education of a generation of young Afghan boys by proto-Taliban and Taliban *madrasa* religious schools in Afghan refugee camps in the 1980s and 1990s, or with Muslim Brotherhood and other nongovernment education providers in Egypt and elsewhere, will be the state's capacity problem: a vacuum in a state education system is a space into which ideological rivals with sufficient resources may move both practically and ideologically, because, as Paolo Freire and others[1] have observed, there is no such thing as an ideologically neutral education system. A technical failure (e.g., of school access promised but not delivered by the state) is readily interpreted as an ideological failure, whether of principle or of will. In our analyses here, as in the world, the relationships between the persuasive power of ideas and the coercive or enabling power of physical resources are complex and bidirectional.

While all these essays analyze fissures and indeterminacies in complex ideological transformation efforts (aka revolutionary movements), as Cati Coe points out, the fragmentation of what was intended is only part of the story: the "failure" of the revolutionary agenda does not mean that nothing has changed. In some cases, the observed failure of the revolutionary model may yield a new model. As William Westerman reports, Myles Horton's observation of the limited effects of a top-down approach to revolutionizing educational process for economically marginal U.S. rural working-class people led to a more profound epistemological shift, in which the potential participants were more effectively recognized by the activists as the generators and holders of knowledge to be propagated through the revolutionary pedagogy. This was an intensification, perhaps a relearning in a new context of N. F. S. Gruntvig's insight in nineteenth-century Denmark, that no knowledge can be functional unless it resonates with the lived experience of the student. The new paradigm was of knowledge emanating *from* the lived experience of the target population, brought into fuller analytic consciousness by the newly dialogic and performance-inclusive educative process. The main U.S. beneficiary of this model was the civil rights movement of the 1960s and 1970s, still an unfinished project whose pedagogical sites shift from generation to generation, hip-hop music perhaps being the latest venue.

In Ghana, the state's attempt to preempt local elders' authority over cultural knowledge and performance for the sake of a national identity project (and for the sake of enhanced top-down political control) seems to falter, as does the state's practical commitment to economic populism in the face of global neoliberal development economics. But more subtly, Coe observed some pedagogical techniques of the schools to be adapted and co-opted into the local economy of cultural knowledge, with as-yet undetermined effects on gerontocratic knowledge institutions and local, hierarchical cultural identity work.

Yücel Demirer's closing observation on the possible future convergence of the revolutionary (Kurdish) and counterrevolutionary (statist) versions of Newroz/Nevruz in Turkey, the potential for cultural and political reconciliation in the further development and enactments of a complex symbol, sounds a hopeful note on the protean qualities of ideological constructions and the creative potential of formal–informal pedagogical interactions. The spring New

Year concept and its accompanying rituals have been repeatedly constructed and reconstructed for ideological and/or devotional purposes, in oral traditions and writings scattered across millennia, the earliest canonized forms perhaps being versions of Zoroastrian liturgical feasts supported by reformist Avestan religious ritual texts from the first or second millennium BCE (the scriptures' dates as well as their meanings are objects of extensive scholarly debate). What is of concern here, of course, is not such hypothesized ancient roots, so prominently asserted in both Kurdish and Turkish state constructions, but the ability of groups to *teach* their claims, to make claims of universality, essential meaning or primordiality "stick" to new interpretations for new social needs, which when acted upon create a new distribution of power.

Within group-articulated formulations and performances, diversities of interpretation, skepticisms, degrees and varieties of engagement by different individuals and subgroups may create dynamic openings (not just "fissures") in the cultural process. Roland Coloma's review of the extraordinary career of Camilo Osias illustrates the power of a superbly gifted individual's enacted reinterpretations of the dominant group's own grounding ideology, to "dis-identify," to intervene and destabilize structures of domination in favor of truly revolutionary new distributions of power to previously disenfranchised group(s). As is the case for all the interventions described in these pages, and many others, the nature of the pedagogical process, over and above propositional content, is one key to propagation and implementation of an ideological intervention in the wider community. For revolutionary change to occur, ideas must be implemented by groups, and groups must be informed and persuaded of the possibility for action, in part by performances modeling that action, which in turn serves to create or consolidate the group itself.

The essays in this volume, overall hardly triumphalist portrayals of either revolution or pedagogy, leave intact some of the mysteries of the persuasive power of enactment. It should come as no surprise that there is no definitive recipe for effective social intervention through education here. In examining from their own various ideological and disciplinary positions some of the indeterminacies, contingencies, and unforeseen consequences of pedagogical efforts in revolutionary environments, this collection's authors have opened a new space for continuing dialogues of both word and action, which in every case examined here, we suggest to be the common, and perhaps most beneficial, effect of pedagogy for social change.

NOTE

1. Brian Street reviews the arguments against technologically driven literacy theory and in favor of ideological theories of literacy developed up to that point, which have remained influential since (Street 1984).

REFERENCE

Street, Brian. 1984. *Literacy in Theory and Practice*. Cambridge: Cambridge University Press.

INDEX